Frederick Thomas Hodgson

The Practical Upholsterer

giving clear directionsFor skillfully performing all kinds of upholsteres'

work in leather, silk, plush, reps, cottons, velvets, and carpetings

Frederick Thomas Hodgson

The Practical Upholsterer
giving clear directionsFor skillfully performing all kinds of upholsteres' work in leather, silk, plush, reps, cottons, velvets, and carpetings

ISBN/EAN: 9783337057831

Printed in Europe, USA, Canada, Australia, Japan

Cover: Foto ©berggeist007 / pixelio.de

More available books at **www.hansebooks.com**

THE

PRACTICAL UPHOLSTERER.

GIVING CLEAR DIRECTIONS FOR SKILLFULLY PERFORMING

ALL KINDS OF UPHOLSTERES' WORK IN LEATHER, SILK, PLUSH, REPS, COTTONS, VELVETS, AND CARPETINGS.

ALSO, FOR STUFFING, EMBOSSING, WELTING, AND COVERING ALL
KINDS OF MATTRESSES, &c.

Rules for Measuring Valances, Curtains, Tapestry, Window and Door
Drapery, Curtain-Rods, Persian Beds, Mantel Drapery, Lambre-
quins, Cushions, Carpets, Floor-Cloths, Festoon Blinds and
Curtain-Hangings, and all kinds of Upholstery.

FULLY ILLUSTRATED

WITH ENGRAVINGS OF TOOLS, SKETCHES OF FURNITURE, AND NUMEROUS
FULL-PAGE ILLUSTRATIONS OF ORIGINAL DECORATIVE DESIGNS.

BY

A PRACTICAL UPHOLSTERER.

NEW YORK:
THE INDUSTRIAL PUBLICATION COMPANY,
1891.

PREFACE.

PERHAPS no branch of literature connected with the furnishing trades has been so much neglected as that relating to upholstering. It is true that books of designs for window drapery, &c., have occasionally been published; but, apart from the fact that they have, as a rule, dealt only with one branch of the upholsterer's work, they have been useless to the apprentice and the young workman by reason of their only showing *what* to do, but affording him no clue as to *how* to do it,— a subject of by no means secondary importance. The want of a work on upholstering of a practical nature has thus long been felt, and many are the inquiries on the subject that have been made from time to time by apprentices and others. These numerous applications for a textbook on the subject have induced the publishers to commission a practical upholsterer of extensive experience to write a fairly comprehensive treatise thereon, which we lay herewith before our readers. To some it may seem rather elementary; but it should be remembered that this very fact will constitute one of its chief recommendations in the eyes of those who are only learners. We, moreover, hold that in works of a technical nature it is impossible to be too explicit; to be practically useful to the largest number they should begin at the very beginning. No previous knowledge on the part of the student should be taken for granted: it being, in our opinion, safer to err on the side of fullness rather than in the direction of omitting rudimentary information because haply it might be familiar to some of the readers.

In this treatise almost every phase of the upholsterer's art has been fully dealt with, and illustrations have been given, with a view to make the text more easily intelligible. It has been written by a gentleman of exceptionally large experience in the upholstering trade, and carefully revised by others equally efficient. We therefore hope and believe that it will prove of great usefulness.

CONTENTS.

CHAPTER I.

HISTORY OF THE ART.

CHAPTER II.

TOOLS AND MATERIALS USED.

CHAPTER III.

UPHOLSTERING A CHAIR.

CHAPTER IV.

PLAIN SEATS.

CHAPTER V.

PARLOR FURNITURE.

CHAPTER VI.

PARLOR FURNITURE (continued).

CONTENTS.

CHAPTER VII.

BEDROOM FURNITURE.

CHAPTER VIII.

BED DRAPERIES.

CHAPTER IX.

BED-HANGINGS.

CHAPTER X.

CARPET-PLANNING.

CHAPTER XI.

CUTTING OF SHADES.

CHAPTER XII.

MISCELLANEOUS HINTS.

PRACTICAL UPHOLSTERY.

CHAPTER I.

HISTORY OF THE ART.

WE have heard it stated that in former times upholsterers were called upholders; and it appears to be not unreasonable when we bear in mind the principal work of upholsterers in olden times, such as the application of tapestry to walls and to articles of furniture.

We have ample proof that the art is most ancient. Monumental records of the very earliest periods in Egypt reveal the fact that the ladies of the City of the Sun were wont to repose on couches and chairs that were stuffed and covered with rich materials of the most costly description; and yet it is possible that they acquired the art from India, the cradle of humanity, and, we might almost say, of all the beautiful arts. It would be futile for us to inquire whence the suggestion of a comfortable seat or carpet emanated. A glance at Nature will suffice to show a common instinct for a soft and warm retreat; and under the rude tents of the inhabitants of bygone ages this instinct would be imperatively exercised.

The Ninevites, the Hebrews, the Greeks, the Tuscans, and the Romans, successively exercised the art of upholstery, as we know from their records; but it is impossible that the materials used could last for ages, hence most of the examples of upholstery are confined to the last few centuries. Perhaps the most luxuriant display of seat coverings and curtains was in the reign of Louis XVI. Many articles produced at this period remain with us, and are being continually reproduced. There are numerous specimens after this style of work to be found in many of the art publications of the past and present day; and with the aid of this work the operative upholsterer in the United States will experience no difficulty in turning out work equal, if not superior, to the furnishings of the days of the luxurious French monarch.

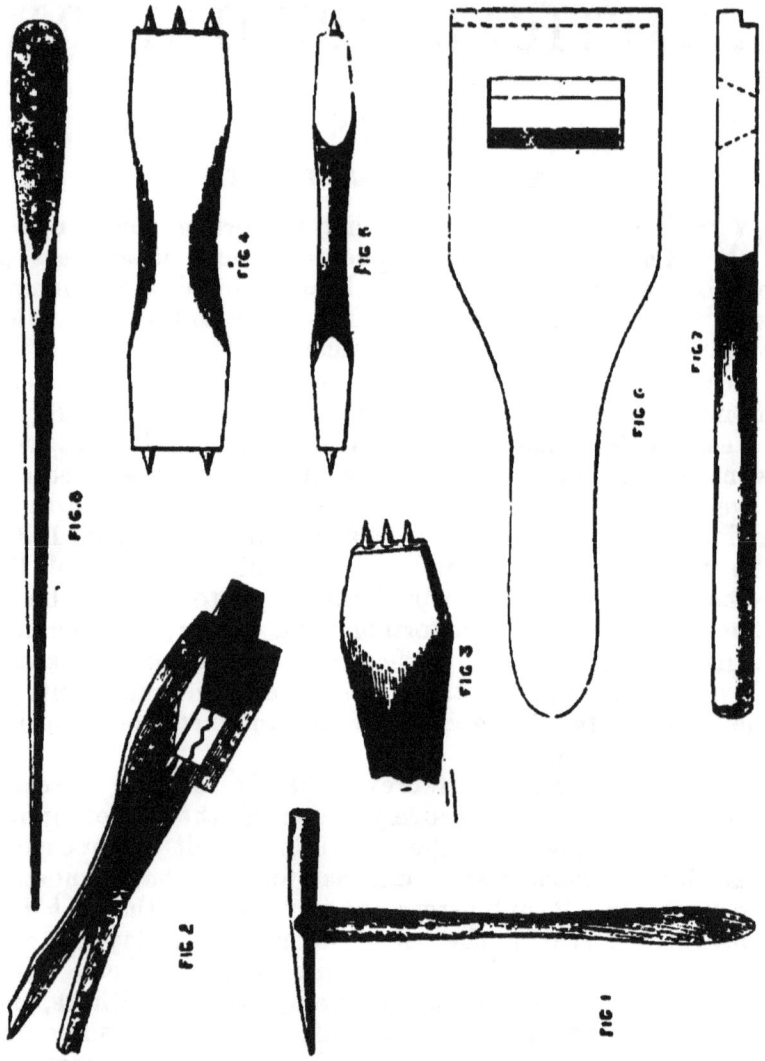

UPHOLSTERERS' TOOLS.

CHAPTER II.

TOOLS AND MATERIALS.

THE tools required by an upholsterer are few in number, the principal being a pair of good hammers (those of French make are the best, and are used by most expert workmen), a web strainer, a heavy pair of scissors for carpets and similar work, a light pair for general use at the bench, a pair of compasses, a strong awl, with the handle in a socket, if possible, in order to bore for driving studs or brass nails, one each of 6 in., 8 in., 10 in., 12 in., and 14 in. double-pointed upholsterers' needles, half a dozen assorted circular needles, a medium-sized screwdriver, a regulator to distribute the hair when tacking down to form edges, a ripping-chisel, a wooden mallet, a 3-ft. rule, a tapeline, and a 16 in. leather bag to hold the whole.

I show some of the tools in the cut on the opposite page, and the following gives a fair description of them:—

1. *Hammer.*—This is much lighter in every way than the ordinary carpenter's hammer, being by comparison a very attenuated affair. A general size is about 13 inches long, with a head of 5 inches from end to end. The thin end is finished with a claw; the other is occasionally roughened to prevent slipping. Fig. 1 gives a good idea of this tool.

2. *Cabriole Hammer.*—This is similar, but with the broad face of the head very much smaller. The workman need hardly provide himself with one of these, as, though a handy tool, he will probably not find much occasion to use it.

3. *Web Pincers.*—The chief peculiarity of these is in the jaws, which are shown in Fig. 2. It will be noticed

that the faces are corrugated or ribbed, in order that a firm grip may be got on the web. The use of this and the other tools will be fully explained in describing work.

4. *Web-Strainer.*— There are several forms of this in use. Two of the most common are here described. Each has its own advocates, and doubtless its own. merits, but they are both good, and whichever is preferred may be adopted. Briefly, it may be said that the spike-strainer is considered by some to be quicker in action than the other, but that it has a tendency to tear the web. Properly used it does not, nor with those accustomed to the other is there much if any saving of time. The form known as the spike-strainer is to be bought; the other, sometimes called, from its shape, the "bat"-strainer, is not, so far as I know, on sale in shops, but is made for or by the user. Both bat and spike-strainer can, however, easily be made, and I therefore describe them fully,— measurements, &c., being taken from two before me that are in daily use in a leading shop. The spike-strainer consists of a piece of oak — any hard wood will do — 7 inches long by 2 inches wide and ¾ inch thick. The center is slightly hollowed on all four sides, and rounded to afford a comfortable hold for the hand; the ends are tapered off to 1⅝ inch by ½ inch. In one end are three round spikes, in the other two. The spikes are ⅜ inch long, or rather they project that distance from the wood, and at the base they are about or a little over ⅛ inch thick. They might easily be made by driving screws in and afterwards filing them to a point, or by sufficiently thick pieces of iron fixed and pointed. A high degree of finish is not necessary. Figs. 3, 4, and 5 show the general shape of this tool, which costs about 50 or 60 cents if bought from a tool-store. Figs. 6 and 7 show the bat-strainer. It is 12 inches long by 3¾ inches wide by ¾ inch thick; one end is rounded and shaped to form a handle, the other is

rebated across to the depth of $\frac{3}{8}$ inch; across the wide part. starting 1$\frac{1}{4}$ inch back from the rebate, is cut a hole measuring 2$\frac{1}{4}$ inches long by 1$\frac{1}{2}$ inch wide on top. In width it tapers down to $\frac{1}{2}$ inch at the bottom, both sides being equally beveled. In length it is the same throughout. All that now has to be done to complete it is to shape a piece of wood so that it will loosely fit the hole. To prevent this small piece being mislaid it is usually attached by a bit of string or strip of leather, or anything convenient, a few inches long, to the bat. In neither of these strainers is it necessary to adhere closely to the sizes, which are merely given to form a reliable guide to those who wish to make their own.

5. *Regulator.*— This is a piece of iron or steel, one end of which is flattened and rounded, the other tapered off to a point. It is made of various lengths — my own is 9 inches — and is sold per inch. Fig. 8 shows its shape.

6. *Needles.*— These may be considered later on as opportunity for using them occurs, when the best form and size for the work on hand will be named. They are made both curved and straight. A couple of each will be quite enough.

There may be some difficulty in obtaining all the above-mentioned articles at an ordinary tool-store, but they can be purchased complete and of good make for about $10, in many first-class stores.

The materials required for ordinary work are: webbing, 5 in. springs for sofa scrolls, 6 in. springs for seats of small chairs, 7 in., 8 in., and 10 in. springs for easy chairs and sofa seats and backs, thick canvas for covering over the springs, scrim for covering the hair for first stuffing, twine, and tacks.

Springs, when they are not compressed, should be from 5 inches to 8 inches for general purposes; but for large work, spring mattresses, &c., they should be from 9 to 10 inches. Springs are made from wire of various gauges, those for backs and soft seats being weaker, as a rule. All springs should be coppered to prevent rust, which soon causes rot and consequent disruption in a chair-seat. The sort having what are known as "coiled ends" are to be preferred to those with tied ends. In the former the ends are fixed by a bend of the end, while in the latter they are bound with thin wire, and are not so good.

Horsehair is the principal material used for filling. It is sold at various prices. The cheapest quality is used for rolls and for very cheap furniture, but it is a mistake to use cheap hair for good furniture. Hair of good quality should be purchased, and it should be teased out by hand. There is a machine made for the purpose, but, according to some, the hair suffers in quality in passing through it. The old-fashioned process of untwisting and teasing keeps the hair in its length: it gets thus better separated, and goes further. Ordinary horsehair-covered furniture is usually filled with hair of a cheap kind for the last stuffing. The very best quality of hair is required for mattresses. Curled hair is undoubtedly the best material, though undoubtedly there is a good deal of rubbish sold as such, but principally composed of various fibers and very inferior hair. It is needless to say no respectable upholsterer would use old hair and pass it off for new. In cleaning and recarding old hair, great loss frequently takes place from waste, and it will be found more satisfactory in the long run to buy new stock. When buying hair, order it loose, that is, carded,—not in the rope.

Wadding or cotton batting is used to prevent the hair-ends from coming through the covering, as well as to ease a slightly harsh feeling of hair alone: it should be laid on the hair soft side downwards.

Alva, or sea-grass, is much used for stuffing and for rolls in very common furniture. Alva is not considered a fiber: it is a seaweed—the *Zostera marina* of Linnæus. In speaking of fibers, we may say that Algerian fiber, or *Crin Végétal*, is a coarse grass found in Algeria. Cocoa fiber is, as every one knows, procured from the inner part of the outer shell of the cocoanut. Mexican fiber is obtained from the leaves of the aloe. Moths do not like alva: this is, perhaps, its best recommendation. Flocks are used for beds, pillows, and bolsters. They are not fit materials for stuffing furniture, and can not be expected to keep their form. They do, however, answer for mattresses and beds, but care should be taken to obtain a good clean quality.

Cocoa Fiber is similar, prepared from the outer husks of the nut. It forms an excellent cheap mattress, and is much used, on account of its good qualities and low price, for hospital and charitable institution bedding.

Feathers may also be used for stuffing. Generally speaking, the more expensive the feathers the better they fill, and a tick filled with good feathers will last for years. They should be thoroughly cleaned and purified.

The old-fashioned idea that feathers may be made fit for bedding purposes by putting them up loosely in bags, and giving them an occasional beating, is an erroneous one. White feathers command a higher price than gray; and though they look better, it is doubtful whether they are more serviceable.

Down is principally used for filling quilts. It is supposed to be eider-down, but very little of this is really used, and it is seldom seen.

The materials used for covering parlor and bedroom furniture are at the present time numerous, but it is perhaps unnecessary to do more than mention here briefly the principal stuffs and their respective widths. These data may prove useful to our readers in estimating the quantities required to cover any particular suite. Tapestries, reps, and damasks are supposed to be one-and-a-half yards wide, but they rarely run to more than 50 inches; Utrecht velvet is 24 inches wide, silk plush 24 inches, cretonne 30 to 36 inches, and satins and silks are made in various widths.

For dining-room and library furniture, moroccos, roans, American cloth, and Utrecht velvet are generally used. Morocco is the skin of the goat, and is by far the best leather used for covering purposes, — its durability and the fastness of its color being qualities not common to any other material. The skins vary in size from 25 inches up to 35 inches and 36 inches in width.

Roans (the skins of sheep) are inferior to morocco, and cost only about half the price: the sizes run from 30 inches to 38 inches wide. They are often used on the outside backs of chairs, &c., when the fronts are covered with morocco, and for that purpose do pretty well; but when used for actual covering they soon become shabby, and should be avoided. Some of the best roans, when quite new, so closely resemble morocco that an experienced man often finds it difficult to decide off-hand which is which. Roans are not so difficult to work as moroccos, being more elastic and supple. Nevertheless, if the suite is to be buttoned down, it is advisable to glue some circular pieces of calico, about the size of a 25-cent piece, on the flesh-side of the skin, and on the spot where the button-twine passes through, as it is possible that otherwise the button will tear through the skin, to the great annoyance of the workman.

American leather, or patent cloth, is about 45 inches in width, and works up similarly to real leather.

Web is a stout linen banding about two inches wide, made up in lengths of 18 yards, &c. It is used as the support of the stuffing in chairs, &c., and should be well stretched, otherwise the seat will drop in the center. Those known as Nos. 10, 12, and 14 are generally used. See Figs. 9 and 10.

Scrim.—This is a coarse thin kind of canvas. The ½ (that is, one yard) is the most useful for general purposes, though

Fig. 9.

in some cases ¾ (or 27 inches) cuts to better advantage. Good makes of this are fairly up to the nominal widths.

Hessian Canvas.—This is like scrim, only coarser and closer. It is generally known simply as Hessian.

Wadding.—Both white and black is a preparation of raw cotton coated on one side with size or thin glue to form a backing. It is made up in bundles of 12 yards in a

piece, and is sold by the yard. The width is about 18 inches;
but as it is split open before using, the available width is
about one yard.

· *Holland.*— Black, white, or bleached, gray or unbleached,
are all used, though the latter two are chiefly employed for
loose covers.

Fig. 10.

Muslin.— Bleached and unbleached.

Twine, for upholstery purposes, is required in three
sorts: a fine thin make for stitching; one rather thicker,
glazed whipcord, for buttoning; and a coarse strong make
packing or laid cord for fastening springs in heavy work.
·

Morocco Leather is the skin of the goat, and is, without doubt, a good material,— though, like most other things, it varies in quality. It is prepared with various surfaces, of which the principal are known as Hardgrain, Crossgrain, and Straightgrain, and may be had in any of them, either dull or bright. Dull is usually preferred, owing to its superior appearance; but this is entirely a matter of taste. Other things being equal, the bright or polished skins wear better, if anything. All varieties can be procured in almost any color. The value of morocco depends not only on quality but on the size of the skins. A large skin is one from which a 3-feet square could be cut, while medium size may be given as about 27 inches by 32 inches. Moroccos are occasionally stamped or embossed with various designs.

Roan Leather is very similar in appearance to morocco, of which it is an imitation,— so much so that it is sometimes difficult to distinguish one from the other without an inspection of the rough side. This is especially the case when new. It is sheepskin, and does not wear nearly so well as morocco, for which it is sometimes substituted.

CHAPTER III.

UPHOLSTERING A CHAIR.

THE first lesson given to an apprentice would be the upholstering of a small chair. We will suppose that such an article is to be completed in morocco, with but-

FIG. 11.

toned seat and welted borders. It is very essential that the workman should know from the first how the article is to be finished: he can then work accordingly.

By an examination of the two chairs shown at Figs. 11 and 12, it will be seen that the upholstering is shallow, or of little thickness. Springs, if used in these chairs, must be very short. Indeed, it is better to stuff such work as this, and complete it without the use of springs.

FIG. 12.

Having cleared the bench, commence operations by giving the chair three lengths of webbing from back to front, and three from side to side, straining each as tight as possible. The webbing (No. 10 or 12) should be tacked with ⅝ inch tacks on the bottom of the frame if springs are used, and on the top of the frame if there are no springs. Supposing the chair to have five 6-inch springs in it, these must be equally and diagonally placed and sewn to the webbing with medium twine, five stitches being used, each equally divided on the ring of the spring. The springs can not be fastened too securely. Now with the lashing-string tie them down to about $4\frac{1}{2}$ inches high, knotting the string to the top ring of the spring, but let them be perfectly upright or they will rattle when sat on. Tack the canvas with $\frac{1}{2}$-inch or $\frac{5}{8}$-inch tacks tightly over the springs, and sew the top rings of the springs to the canvas with five stitches equally divided. Knot each stitch separately. This will be best done with a crooked needle.

For the first stuffing of hair, run a twine round the edge of the seat to hold on the hair; pick or string on a fair body of hair (not much in the middle, as the chair is to be buttoned); place the scrim over the hair, keeping the thread or bridle quite square with the chair, as it makes the work much easier; tack the scrim temporarily all round into place; stitch it to the canvas on the springs with a double-pointed needle, and let the stitches be about three inches long and 4 inches from the outside edge. Pulling the scrim down to the canvas in the middle of the seat by thus stitching it thereto will necessitate the outer edge of scrim to be lifted and filled up, to form the outside edge of seat, and this must be done without disturbing the center.

Thus far the work has been tolerably easy: we now come to the more difficult portion. Fill in the outer edge firmly, and bear in mind that when the scrim has been tacked down with $\frac{1}{2}$-inch tacks and stitched, as the middle of the seat has been, it should be about 3 inches or $3\frac{1}{2}$ inches from top of molding, the perfect shape of the chair, with the edge of seat slightly hanging over. Three rows of stitching are sufficient, and the first stuffing is then completed.

Now commence the second stuffing. Draw a line down the center of the seat from back to front and mark the places for ten buttons,— three in the front row, two in the

second, three in the third, and finishing with two in the
back row. The buttons should not come nearer than 3 in.
to the edges of the seat. Make a small hole in the scrim

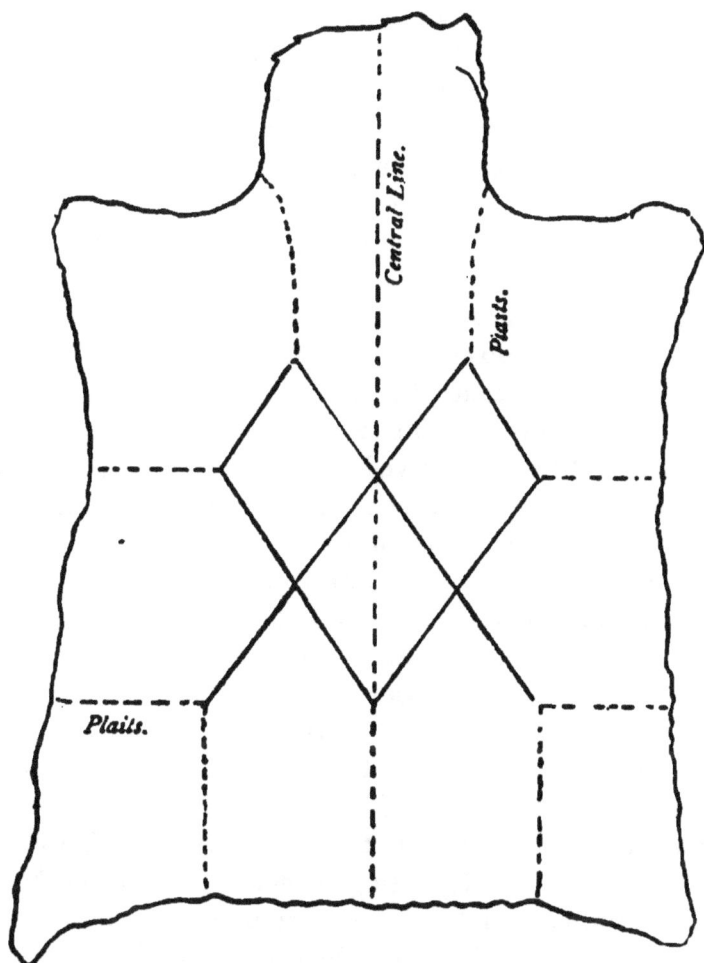

FIG. 13.

with the scissors in each place marked for the buttons, so
that the right place for each may be felt when the hair is
picked on for the second stuffing.

The skin must now be marked and creased. If the

buttons are required very deep, 2½ inches may be allowed
for fullness: more usually 1¼ inches is allowed to each
. diamond. The neck of the skin is placed to the back, and
is marked as shown in sketch (Fig. 13). The remaining
skins can be marked by this one if more than one has to be
done; and should there be any marks in the skin, fit them
in the plaits so as to hide them. Pick on the hair for second
stuffing. Care and judgment will have to be exercised in
doing this, in order that the leather may be filled out firmly,
and as clear of creases as possible. Next put the sheet or
pound wadding on, then the skin, and after this put in the
buttons with a slipknot, using button twine, and pulling
them about half down. Care must be taken to work the
hair away from under the buttons, and then pull them
down to their places. The twine must then be knotted and
cut off, and the plaits worked out clean. Wherever the sides
require more hair fill them in, and pin the skin to the edge
of first stuffing, working all the fullness into the outside
plaits which are square with the seat. Now cut the skin off
to the exact size, allowing about three-eighths of an inch
for turnings; put a stitch or two in the plaits on the outside
edges to hold them in place and to avoid any fullness.

The strip which is cut off will come in for the border
and welt: in the welt use the lashing twine. The border
will require two joints, and these should come on the sides
and about half an inch from the front corners. The following is the process of jointing or splicing: Cut the border
quite straight; chamfer the ends to be joined so that they
will lap over each other about half an inch. This must be
done with a sharp knife. Some upholsterers make use of
curriers' paste to secure the joint, but glue which is nearly
cold is to be preferred: if used hot it will penetrate the skin,
leaving a black mark even when dry; and it is well to bear
in mind that if any gets on the face of the leather it will

disfigure it. Paste will also leave a mark if it be allowed to penetrate the leather, but the glue when cool is quite safe. When the joints are dry, strain the border tight round the chair and pin it temporarily on; then snip small notches in the seat and border, so that they will correspond and be a guide to the upholsterer: there should be three notches on each side, and the same number in front. Cut and join the strip for the welt in the same way as the border, and it will then be ready for the upholsterer to sew. When the border and strip are stitched, turn them up and sew to the edge of the leather seat.

Tack a small piece of buckram on each corner of the seat in the front. This will help to preserve a good form. Next put some wadding under the border and tack it with ⅜-inch tacks, leaving the welt straight and the border without a crease. If the instructions here given have been carefully carried out, the chair should look perfect. If it is to be finished with banding and studs, drive the studs in about 1½ inches apart; but if the chair is to be close studded, no banding will be required. Use canvas on the bottom only when springs are employed. Two pounds and a half of hair is the average quantity used, whether the chair has a spring seat or not.

It must be a matter of consideration for the upholsterer whether the extra work entailed by welting is compensated by any advantage. If he thinks it is, this is how he must go about the work, during which it will be well to enlist the services of some one who can sew with an ordinary needle and thread, besides being able to use a thimble. Sewing through four thicknesses of morocco is not exactly easy work, so it will not do to put anything down that is not quite flattering to the exalted abilities of those beings who, as somebody—is it Scott?—tells us are "uncertain, coy, and hard to please." But, after all, it is your friend, not ours,

who is going to do the sewing, so what does it matter?
You can quote Scott, too, and prove her to be a ministering
angel, as he says she is, "when pain and anguish wring the
brow." They are pretty certain to if you, unaccustomed,
use a thimble.

Before beginning a welted-edge seat, it may be desirable
to know that this form is hardly so durable as a plain-edge
seat, if subjected to constant wear. The mode of stuffing is
the same as that described for the other seats with stitched
roll,— the only difference, if difference it can be called, for
it is only comparative, is that the upholstery should be, if
anything, firmer, especially on the edge, and the hair spread
very evenly. Unless these two points are attended to, the
seat will not keep its shape long, or rather the wear on the
edge will be more conspicuous from the welt emphasizing,
as it were, any unevenness or irregularity.

Let us suppose the seat is to be plain on top, without
buttons, as it is a simpler piece of work, and any additional
instructions can be given after it has been described. The
skin, or a piece of it sufficiently large to cover the top of the
seat, is laid on the stuffing and fastened by a few pins driven
through it into the stuffing. There is a special kind of pin,
or small skewer, sometimes used for the purpose, but a good
sized pin will do as well as anything: in fact, almost any
piece of wire three or four inches long will serve, if it can
be driven through the leather. Those long steel pins with
a fancy knob at the end, which ladies use to fasten the hair
they buy (I do not know what the present fashionable
make-up is called, but I dare say if you ask any young lady
she will tell you) on to their heads, are just the very thing.
Mentioning this reminds me that the pin is used just in the
same way to fix the morocco down that the said hairpin is
when devoted to its original intention. It is simply thrust
through into the hair, just as a nail would be into the frame,

The pins should be driven through close to the edge of the morocco, which should be a trifle larger than the seat of the chair, that it may overhang a little when smoothed down, and stretched tightly but not excessively. Two or three pins on each edge will be sufficient, as they are only needed to keep the covering in position whilst it is being marked to the exact size it is being cut down to. The marking is done by drawing a line with a piece of chalk all round the skin just on the edge of the chair, so that when the margin is cut off the top of the seat shall be just covered, and no more, by the leather. We have omitted to say that notches must be cut, when pinning, for the back legs, as already directed.

Now cut the bordering, which is best if in two pieces only, one of them being for the back and the other for the sides and front. If a sufficient length for the latter is not convenient, it may be made in three pieces, which should be stitched together so that the join just comes on the corners, or the pieces may' be skivered and fastened with strong paste. A good adhesive medium is made by adding a little resin to ordinary flour paste.

Skivering, as it is called, is done by shaving away, or beveling, with a knife, the two pieces where they are to join. One piece is beveled on the face, the other on the back of the skin, and joined by putting the latter over the former. Strong paste is necessary, and, as already mentioned, when giving directions about buttoned seats, care should be taken that it does not stain through. A well-made joint of this kind almost escapes notice, but, on the whole, we prefer the sewn corners. When cutting the borders the top edge should be cut evenly: the lower one is not so important, as it can be trimmed off afterwards.

The border should have at least $\frac{3}{4}$-inch fullness in the width. Much more would be waste, but less can hardly be done with. Both back and side borders must have an extra length or fullness of not less than half an inch to allow for turning under at the back legs. Pin the bordering on to the chair temporarily, stretching it fully. Make notches on it and on the covering when in position, so that on sewing the two together they may fit rightly. The notches, of course, must not come beyond the edges which are turned in, otherwise they would disfigure the seat: one on each side and another on the front will be sufficient. Another piece is required for the welt. This should be a strip, say ¾-inch wide, and as long as the bordering. It should be in one length, that is, for the front and sides; and if a join has to be made, it should be done by skivering, not sewing. A piece of string, in thickness varying according to the desired bulk of the welt, will also be necessary to serve as what may be called a core for the welting leather. Ordinary whipcord, or stitching twine, is as useful as anything for this purpose. We do not know whether it is necessary to explain for the benefit of any who may not understand what is meant by a welted seat; but as it is difficult to convey the correct impression, we may refer to an ordinary mattress as being the most familiar example. True, all mattresses have not welted edges, but many—we may say most of the good ones—do. On looking at it, see if between the cover on top and the border there is a small strip, rounded, sewn into the edge. If there is it is made with a welt, and a similar style is just what is wanted in the chair. To sew the welt, border, and top together, after folding the welt-strip, put the pieces in their proper order with their edges even. The top covering will have its right side uppermost, the welt folded with the string shown by the black dot within it, the bordering above it with its right or surface side downwards. The edges are now to be sewn together by stitching or "whipping" round them with strong needle and thread. And here the advantage of the notches will be felt, as without them it would be difficult throughout the whole length to preserve equal tension of the top and the border. With them a kind of check is given (about) at every third of the length. If any difficulty, even with three of them, be experienced, there is no reason why several more should not be made; but in practice, three notches are generally found to be sufficient. It is not convenient always to make a notch; so if the work is being covered in a material which it is not judged advisable to cut, a chalk mark, or, indeed, any mark, such as a stitch, may be made instead.

No great care is necessary in doing this part of the work, as the stitches need not be very regular nor close together, a quarter to a half inch being near enough. When this sewing has been done, another row of needle-and-thread work, requiring more regularity and fineness in sewing, must be put in. Good strong thread must be used here, and if it is waxed it will be the better. The stitches must go through the four thicknesses as close to the cord as possible, and the row be made up of "backstitching." We think this is a common term, not peculiar to upholstery work, so it need not be explained further. When the stitching has been done, the cover may be put on the seat and sewn to it. This sewing may be, and often is, omitted; but it will be hardly advisable for the amateur to do without it, even though an experienced upholsterer occasionally may dispense with it. Then, again, coarse sewing will do as well as fine, the great matter being not to distort the straight even line of the welt. The stitches are merely to attach the covering to the seat; and as they are close to the welt, being run through the fourfold edge into the seat, it is obvious that when the chair is in use the covering can not so easily be displaced as it otherwise would be. When sewing the covering down see that it is well and evenly stretched, and that the welt is straight from end to end. If it should not prove to be so when the sewing is finished it will be better to undo and sew again, as it would never do to allow the cover to remain with the welt on the edge of the seat in one place, and half an inch over the top or down the border in others.

All that now remains to be done is to turn the border down and fasten in the usual way, finishing off with gimp, banding, &c., as already described. The border ought to be drawn tightly down, otherwise it will soon look puffy and loose. To prevent this appearance, it is not at all a bad plan when, or rather before, covering a welted seat, to pull down the stuffing by a few ties in order to flatten it. The ties should be fastened underneath the seat, that they may be removed when the covering is on. The stuffing will then spring up to its normal state, with the result of a well distended and tightly stretched covering.

When buttoned seats are welted, the proceedings are altered to a small extent, as the border and welt are sewn to the top of the covering after this has been attached to the chair by the buttoning, and the pleats have been formed. The buttoning is done as already described, after which the covering is trimmed to the top of the seat; the welt and border are then sewn on. Before this, however, in order to prevent the pleats from the buttons to the sides, back, and front, from being disarranged while sewing on the welt, it is always a safe plan just to stitch them at the ends. This prevents them unfolding during the subsequent sewing, and insures each being kept on. Perhaps we ought to say that the pleating is confined to the top of the covering, it being stopped by the welt, so that the border is plain and without any pleats.

Corded Edges.— Corded edges, we have already said, are principally used on drawing-room and other fancy chairs

when the coverings are of two different materials, as, for
example, a tapestry top and plush borders. The appearance
of the edges is very much the same as in the class of coverings which have just been treated of, only in place of the
welt there is a fancy cord either of one color or a combination of the colors in the covering. The mode of working is,
in general, the same as for welted edges, though there is
sufficient difference to justify a separate explanation without
going any very great length into details.

To begin: after the stuffing has been done, cut the cover,
that is, the part for the top of the seat, about an inch larger
than it will show when finished. Place it on the seat and
fasten temporarily to the back edge with a couple or three
pins of the kind already named, and smooth it over to the
front, and fasten in a similar manner, after which do the
same on both sides, stretching tightly. Cut at the back
corners for the legs close up, as a soft thin covering is more
easily turned in. Only experience, however, will teach exactly the relative hardness or softness of coverings; but as
some guide it may be stated that plush is a soft and morocco
a hard covering,— one of the hardest, indeed, in general use.
if we except pigskin, though this can hardly be called an
ordinary material, though it may become popular. In our
opinion it is a most desirable covering, and free from the
objections that have been often justly urged against pigskin
for upholstery purposes. On account of its comparative hardness, however, it is not a material so suitable for a beginner
to commence upholstery with as something softer.

The cover being pinned on, cut it evenly round, leaving
a fullness of about half an inch outside the pins. Turn the
edges under, putting more pins in as the work proceeds,
say 1½ inches distant from each other, so that the cover
may be kept well in place. The border is then stitched on
with strong thread,— that used in making carpets does very

well. Ordinary sewing-cotton is not strong enough. The covering is then sewn to the edges of the seat as in welting, the pins being removed as the sewing advances. The bordering may then be tacked down to the frame. To complete the edges, the cording must be sewn on just along the join. Tie round the cord to prevent it fraying out, as directed for gimp, and start at the back, preferably near the leg. In bringing the cord past this to the side of the chair, do not stitch it on the top, but press it down between the stuffing and the wood, so that it is not seen. Continue the sewing on the side, as close to the leg as convenient, and fasten it round the seat till the starting-point is arrived at. We ought to have said, when beginning to cord, do not sew quite at its end, but leave half an inch or so loose, and cut off the other end in the same way when the stitching is finished. These two ends are then stuffed through the seam between the top and the border, so that the join may be as little perceptible as possible. It occasionally may happen in cutting the top of the covering that there is sufficient to hang over as far as the wood of the frame to which it may be nailed, instead of being cut close to size and pinned. In such a case, for instance, where two seats can be got out of a width of material with a few inches to spare, but still not sufficient to use for any other purpose,— in fact, waste,— no object would be gained by cutting the covering exactly to the top. We merely throw out this hint for what it is worth, as there is no fixed rule, and can be none, where so much depends on the material which is used. It is in such cases as this that the *art* of the upholsterer comes in.

CHAPTER IV.

PLAIN SEATS.

CHAIRS with plain seats and welted borders are proceeded with exactly in the manner described in the last chapter, as far as the first stuffing is concerned, except that the springs may be left a little higher. Some workmen first finish the plain seat in calico, and then strain the skin over it, and cut it to shape on the calico. This is a difficult, and not very safe, course to adopt. It is much easier to cut the skin on the first stuffing. In order to do so, lay the skin clean out; do not strain it; pin it round, allow for turning in, and cut to shape. Prepare the border and welt as directed in the previous chapter for a buttoned seat.

For an unbuttoned seat, strain the border round, and notch both seat and border for the guidance of the upholsterer; take the leather off, and, when it is stitched, hammer the welt flat. The strain on the border will prevent all wrinkles, and will bind the seat to shape.

The reader will thus see that the seat is all prepared before the second stuffing is attempted. Now pick on the hair and finish it in calico, stitch ties up through the webs, springs, and seat, and pull all down flat, making knots on the under side of the webbing. These ties are merely temporary, and will be afterwards cut and drawn out. The seat being tied down, the skin can now be drawn on easily; stitch the welt to edge of the first stuffing, and tack down the border to its place. The temporary ties may now be cut and drawn out, and the seat will then rise up as tight as a drum, and be in good form. Any number of skins can be cut from one pattern, taking care always to notch border and seat. They can not but be correct; but remember to strain all the borders and pattern lengthways together on the board, notching them to the pattern.

The above directions answer for any shape. Welted borders are not very often ordered, except for first-class articles. If welting be not required, the chair would still in other particulars be forwarded in a similar way; the skin, whether buttoned or plain, would be drawn over and tacked down to the seat-rail molding and finished as before directed. All plaits can be got out, if sides and front are straight, by temporary tacking and shrifting till you have eased them out. If round shaped, the plaits can not be removed entirely, but let them be upright.

Easy Chairs in Leather.—There are no fixed rules as to the heights of springs or edges in easy chairs, for scarcely two are alike in this particular. The style of the chair must be taken into consideration, and the stuffing should be adjusted with due regard to the proportion of the article. If it be a large easy chair, there should be a bold swell on the lower part of the back, and high edges on the seat; but if the chair be small, these parts must be proportionately lower. If the seat is to be welted, the instructions before given for a small chair will be applicable. In the allowance for fullness in the diamonds, allow $1\frac{1}{2}$ inches or $1\frac{3}{4}$ inches for each diamond,— that is, if the diamonds on the scrim are 7 inches long by 5 inches wide, then mark the skin $8\frac{3}{4}$ inches by $7\frac{3}{4}$ inches. Gentlemen's or ladies' easy chairs would be alike in this respect.

Lounges and Settees.—Lounges and settees are considered more difficult to upholster, as the skins have to be joined to make them large enough. The seat of a lounge usually takes three skins, and sometimes part of a fourth. Moroccos are seldom large enough to tack down when the seat is buttoned. If the seat be plain, cut them perfectly square across the seat, and join them with a small welt;

then pin them over the seat on the first stuffing, and cut them to shape. Strain the border well round to avoid fullness when finished, and join it up to the seat according to the instructions already given for a small chair. If the scroll and pad at the back are plain, proceed in a similar manner. If the lounge is to be buttoned, first mark the places for tufts on the seat on the first stuffing. Let the first row of tufts be 3½ inches from the front edge of the lounge, with diamonds 7 inches across the seat by 5½ inches lengthways. Cut small holes in the scrim where the tufts come, as before directed, and mark the skins as follows. The allowance for fullness on a lounge-seat is 2 inches across and 1¾ inches in the length for each diamond: this will be marked on the skin 9 inches by 7¼ inches. The reason for allowing more fullness across the seat is that, being on the round, the seat takes a large sweep, while lengthways it is straight. Mark a piece of paper right out to the edge with as many diamonds of the size given as a skin will contain. Place the skin face up on the board, and the neck to back of the seat; put the marked paper on the skin, placing it in the most economical position. This will be ascertained by measuring from the button-marks nearest the edges, to see if there be leather enough to cover. The best position having been obtained, prick through the tuft-marks on the paper, and thus mark the skin by it. Should the skin be small from back to front, piece up at the back, and mark other skins by the same paper until the seat is completed. The joins must run zigzag in the plaits, as shown in the accompanying illustration (Fig. 14). The button-marks should come exactly together, and be sewn through, not over. Next welt and border according to instructions previously given for small chair.

If a lounge-scroll is buttoned, it will seldom take less than a skin and a half to cover it. In marking the diamonds

FIG. 14.

on the scrim for the scroll, commence 8 inches from the seat, and mark them 7 inches across and 5 inches lengthways, as on the seat, then run them up over the scroll as far as they will go. Allow 1½ inches for fullness of each diamond across the scroll; and here note particularly that the allowance for fullness lengthways of the scroll must be increased at every button from the lowest button line one inch for every diamond; for instance, if there are four diamonds up the scroll, the allowance on the first will be 1½ inches, on the second 2½ inches, on the third 3½ inches, and on the fourth 4 inches. These allowances will be safe for any ordinary scroll, but it may be varied a little according to the quickness of the sweep. The reason for the increase of the allowance in the diamonds of the scroll is obvious, inasmuch as when the scroll is stuffed the quick curve of the scroll is still more quickened by the additional sweep (fullness) of each diamond, and is necessarily much larger than the surface of the first stuffing. It is undoubtedly one of the most difficult tasks for an upholsterer to make a good-shaped scroll. On the following page is given two illustrations (Figs. 15 and 16) exemplifying the manner of upholstering lounge-scrolls.

In stuffing the back, allow about 1½ inches in each diamond for fullness. Do not place the bottom row of buttons on the back quite as low down as on the scroll, so that the tendency to wrinkle across the diamond may be prevented, and the shape of the swell improved. If there be an armpad, the buttons should be about 5 inches apart; small pieces can be used up on this, only see to it that the joints come in the plaits; 10-inch leather will be about wide enough for the arm-pad,— the buttons are held down by means of twine passed through the center of the pad, and held down on each side by tacks,

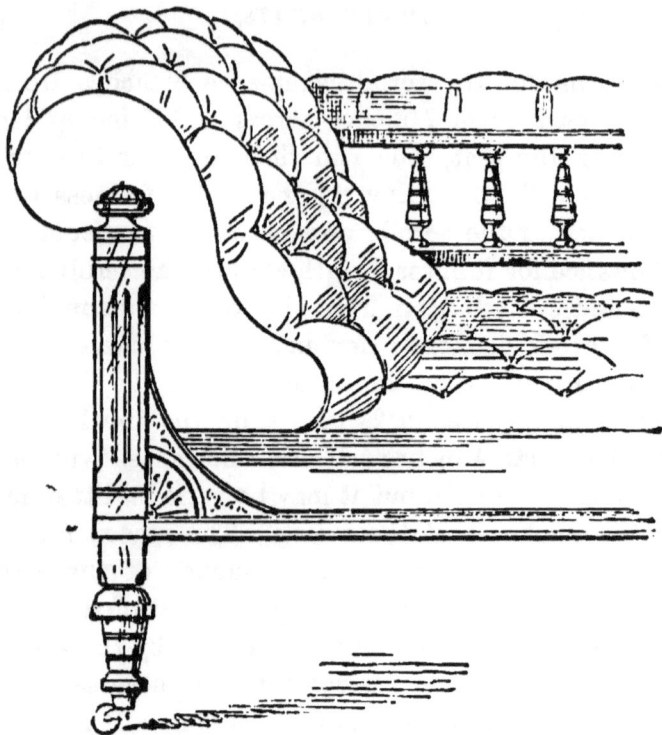

FIG. 15.

FIG. 16.

American oilcloth is treated in a manner similar to leather, and the same allowances for fullness will answer very well.

Hair seating can be had in almost any width, but can only be used on plain seats. It should be laid rather loosely on backs and scrolls, since the buttons will pull it to shape. Haircloth can not be plaited. In order to upholster an article in haircloth, properly finish it right out in black holland or canvas: if white material is used it will show through the seating when finished. The seats are usually welted: this enables the workman to use narrower widths, and these are cheaper in proportion than the wider ones. Having cut out the seating and borders to size and shape required, strain the border round easy and welt it. When the hair seating is placed on and tacked a good many wrinkles and plaits may appear in it. These are difficult to pull out, but they can be removed by moistening with water. When the cloth gets dry they will be quite gone. This moistening process saves much labor, and is not known to every upholsterer. The writer has often noticed common furniture upholstered in hair seating, and long wondered how the plaits, &c., had been got rid of, and the fabric made to look so smooth, until he learned the moistening process, which will be found to answer admirably.

CHAPTER V.

PARLOR FURNITURE.

THE upholstering of parlor furniture in soft materials is easy work when compared with the covering of chairs, &c., in leather. The edges of the seats need not be made anything like so hard as where leather or hair seating is used, but should be kept soft. The springs also should be more pliable.

Supposing a suite is to be upholstered in tapestry or rep, the scroll should then be furnished with soft 5-inch scroll springs. Tapestries and reps are wide materials. They are supposed to be 54 inches wide, but are scarcely more than 50 inches in width. There will be no occasion to join such fabrics. If there be a running pattern we would prefer to see it running along the seat and over the scroll lengthways, and up the back. If the material be wide, and the above course adopted, no joining will be required in the seat. If the material be narrow, as is the case with velvet (24 inches), it will be better to have the joint lengthways, putting it to the back, than to have two joints across.

Spring-edges.—In making a spring-edge put 8 or 9-inch soft springs in the middle, keeping them farther from the rails than when not spring-edged, so as to give the edge-springs room. Lash them in place, allowing them to stand as high as possible, consistent with the proportion of the article of furniture. Put the canvas over and tack it to the top of the rails on the extreme inside edge. To form the edge use soft 6-inch springs; fix them firmly upright on the rails, tie them all to the same height; the string that holds them would be best knotted to the top ring of the spring and held down on each side by ⅝-inch tacks. To form the shape of the edge, use spring wire and bend it to the exact shape of the rail, tying it tight to the top ring of the spring-edge; cover them with canvas, which latter should be sewn to the canvas already over the seat, and about three inches from the top level, and to the spring-edge. This will allow the springs in the seat and edge to work independent of each other and without strain. First stuff soft and free; the stitched edge should be bold and overhanging, and finished on the wire edge. See that the springs do not clatter; sew a strip of canvas to the wire edge, which must be tacked to the seat-rail, regulating the height of border by pulling to required shape. The second stuffing will look best finished exactly under the roll with a bold cord, one row of buttons on the border or frilled.

Plain Seats.— Suites are often upholstered with plain seats and buttoned backs and scrolls. These can be bordered and finished with cord on the edges, or, if it be a cheap job, they may be tacked right down to the molding. For ordinary materials the plain seats should be first stuffed in calico; but if the article is to be covered in velvet or plush, it should be finished throughout in calico, and covered with the velvet or plush afterwards.

Buttoned Seat.— Mark out the scrim for buttons as explained in previous chapters. In measuring the sizes, allow the fullness for seat in addition: for a lounge or sofa-seat, $2\frac{1}{4}$ inches across and 2 inches lengthways for each diamond. This will do for almost any parlor furniture or soft covering. The allowance on the back of the lounge or sofa should be $1\frac{1}{2}$ inches each way; on the scroll $1\frac{3}{4}$ inches across, increasing 1 inch in the length of fullness for every diamond from the lowest row. Scrolls are often, especially in high-class work, buttoned down low to the seat.

For the seat of an easy-chair, whether a lady's or gentleman's, allow for fullness 2 inches each way, and for the backs $1\frac{1}{2}$ inches. On small chairs, $1\frac{1}{2}$ inches each way for fullness should be allowed. If the suite is to be bordered, turn the cover under and sew it to the edge of the scrim on the top. Some upholsterers stitch it first in under the roll, but this is not to be recommended, and for the following reason: When the cover is sewn under the roll, and the top of the seat has had a little wear, it gets loose and bags down, making the work look unsightly; whereas, if the cover is sewn to the top edge of the scrim, it can not possibly move.

The border is often varied in color, plush borders being plentifully introduced at the present time, but the seat and the border should either match well or form a happy contrast. Where there is no molding round the seat-rail the addition of a fringe often has a pleasing effect. These instructions for the upholstering of a suite apply to almost any kind of sofa, easy-chair, or small chair.

Quantities.— The quantity of covering materials for an ordinary parlor suite with moldings on the seat-rails and plain seats would be as follows:—

Velvet or plush, ...	18 to 20 yards.
Rep, ...	8 " 9 "
Tapestry, ...	8 " 9 "
Cretonne, ...	14 " 16 "
Cord, ...	22 yards.
Gimp, ...	36 "
Buttons, ...	½ gross.

For upholstering a similar suite with buttoned seats and backs, a much larger quantity of material would be required, namely:—

Velvet or plush, ...	24 to 26 yards.
Rep, ...	10 " 12 "
Tapestry, ...	10 " 12 "
Cretonne, ...	18 " 20 "
Buttons, ...	2 gross.
Cord, ...	22 yards.
Gimp, ...	36 "

French Work.— There has been a decided revolution in upholstery work during the past few years. Artistic forms, combined with French luxuriousness, are much sought after, and some of the designs thus introduced are decidedly tasteful. This style of work is upholstered very soft, and

Fig. 17.

can only be done with a good quality of hair, otherwise it will lose its proper shape before it has been in use any length of time. The accompanying illustration (Fig. 17) illustrates a pattern of this class of furniture, which is just now in favor.

The seat of such a chair should be filled with very soft 8-inch springs. There is no stitched-up edge to the seat, but the scrim is tacked down to the bottom of the frame in front, and is finished with a round edge in calico hanging slightly over. Having upholstered the plain part of the seat in the usual way in tapestry, stitch it to a line previously marked on the calico. Recollecting that the front is tufted, mark the half diamonds on the calico round the front, as shown in the illustration, keeping the buttons about three inches apart, and snip holes for button-marks. Next mark the plush; and, in doing so, allow a good $1\frac{1}{2}$ in. for fullness. Fill in hair on the top of the calico, and tuft it round, sewing it to the tapestry so as to keep it in place, and finish with cord or gimp, which will cover the stitches. The tacks on the rail are covered with festoons. These should also be of plush, and finished with a fringe about 1 inch or $1\frac{1}{2}$ inches deep. The inside of the back is webbed, canvased, and finished in tapestry, without any stitching. The sides are done similarly. The pad which runs round the arms and back is finished in scrim cut on the cross: this makes it softer, and helps it to hold better than if cut square. It is necessary to use good hair for such an article, and to tack it rather firm on the pads; but it should not be stitched, except in the front scrolls. Mark the pad for the buttons at about four inches distance, and snip holes to allow the buttons to sink a little. Lay on the wadding, and cover with the plush, finishing with a good-sized cord. The whole of the chairs will thus be of tapestry, with the exception of the buttoned part, the end arm-borders, and the festoons. The festoons are cut wider at the bottom than the top, the bottoms being rounded a little. This forms the fullness and the plaits near the bottom, leaving the top gradually plain. These are made and tacked on separately, a bold cord covering the tacks as on the inside back.

The next illustration (Fig. 18) shows a similar chair. Our object in giving it is to point out the bolster on the top of

the back. It looks at first glance as if it were made with
the frame; but this is not the case. The bolster on the top
is formed entirely in the stuffing, and is made as follows:—
Pick or string on the top rail a good body of first-class hair,
and make it quite firm, but not tight; cut the scrim on the

FIG. 18.

cross and about 20 inches wide; tack it down to form a well-
shaped bolster, and tack the ends down; stitch it up to a
fine edge, and make it of an easy and a graceful contour, as
shown in the illustration. There is no stitching required in
any other part of the bolster, nor in any other part of the
chair.

Needlework is often placed in the hands of the upholsterer, to be used for coverings. Work of this kind is frequently out of the square, and puckered. This may be rectified by straining the work on the board as tight as possible, and quite square, with the face downwards. Having placed a clean cloth under it, now damp and press it until quite dry. When taken off the board it will be found perfectly straight and square.

FIG. 19.

Crewel-work is generally wrought on satin or cloth, and may therefore be suitably bordered with plush of a shade similar to that used in the needlework. Wool-work appears to advantage with black cloth, the cord and gimp being similar in color to the wool employed.

Spanish Chair in Needlework.—There is, perhaps, no chair more suitable for the display of strips of needlework than the kind here illustrated (Fig. 19). It is also inexpensive, being entirely stuffed over. Suppose the workman has one to do with needlework center and plush sides and border as per sketch shown. He will then first stuff it without springs, keeping it quite flat across, and giving it a good

side-line, as shown. Mark the middle for the strip of work;
pick or string on a little hair and wadding; put on the
needlework; pin it to a perfect line; sew it in its place with
a 6-inch needle right through the canvas and webs. By
letting the needle slant outwards from the work a more
rigid stay will be obtained. All chairs this shape, when
upholstered plain, have a tendency to wrinkle across, and
this can only be avoided by keeping the covering quite flat.
On the buttoned side-margin it is sufficient to mark the inside line of buttons, allowing for fullness 1½ inches where
it is pretty straight, increasing a quarter inch to a half inch
where it rounds (top and front), decreasing the same amount
in the hollow. A cord partaking of the colors of the work
and plush hides the point. The borders should be finished
with the same. Too much care can not be given to the outlines, as on this depends the appearance of the whole. We
can safely say that if the work as here described be carefully
carried out there is no chair that sets off needlework to
greater advantage.

CHAPTER VI.

PARLOR FURNITURE.—(*Continued.*)

French Settee.—The settee illustrated in Fig. 20 is most suitable for upholstering in plush and tapestry, or any two colors to harmonize nicely. Let the springs in the seat stand high and soft. The front edge should have no stitching, but be simply filled in and rounded gradually off to the rail. On the back and end of the arms stitch a well-shaped edge, as shown in the illustration; finish seat and back in calico, and mark on it the shape to which to work the borders; also set out the button-marks, snipping holes to allow them to sink. The plain portion of this settee is supposed to be covered with tapestry, and the buttoned borders and festoons are of plush. The base festoons should be cut rounded on the bottom, and about three inches wider than the top, each one to be separate and finished before being tacked on, as is also the plush festoon shown on the top of the back. This is really added after the article is completed, and simply hangs loosely over, being finished with narrow fringe. On the front border a narrow bordering of hair should be picked or strung on, to fill out the diamonds pretty firm, as this part and the arms are exposed to the most wear. Allow $1\frac{1}{2}$ inches to $1\frac{3}{4}$ inches to the diamond for fullness. The fringe under the base festoons could be dispensed with; or tapestry, the same as used for the seat, might be substituted.

Circular Ottoman.—The top part of the box of the ottoman shown in Fig. 21 should be made of hard wood, not less than one-inch stuff, framed together so as to admit of the upholsterer's needle going through when buttoning the border. In the top five 6-inch springs could be placed. Tie

them in their place,—of course not too high, as the article
when complete should not, including castors, be more than

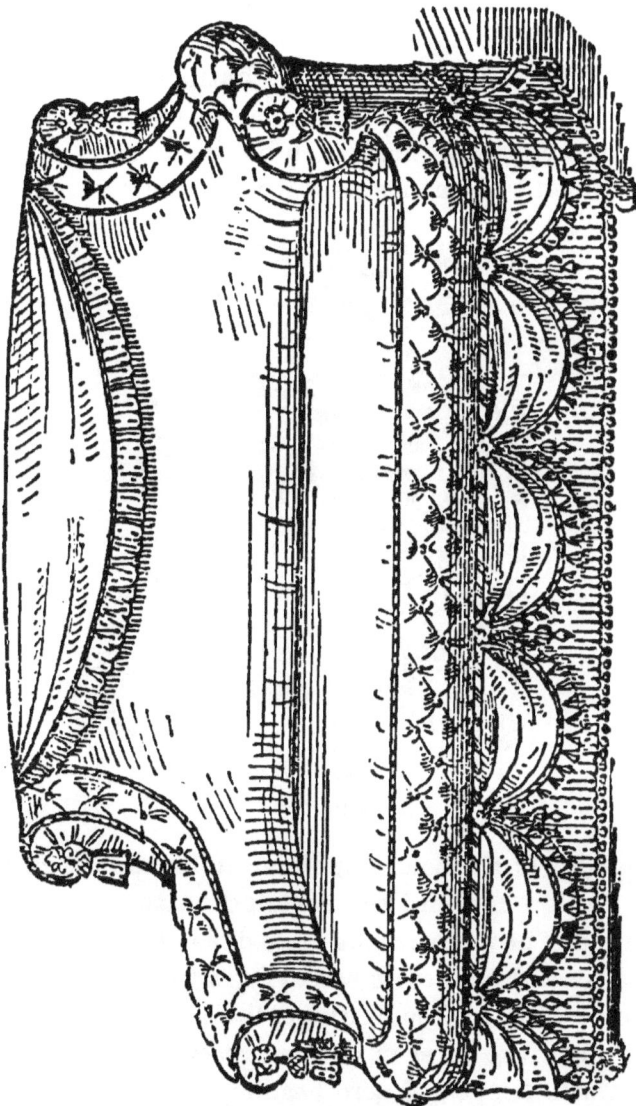

FIG. 20.

18 inches high. First stuff it soft, no stitching being employed; finish it in calico rounded gradually down to the

rail. Mark the top circle for the border to finish on the
calico; also set out the tufts on the calico. On the border
allow $1\frac{1}{2}$ inches down by $2\frac{1}{2}$ inches round the seat for fullness;
put on the top circular piece, sewing it to the calico; now
snip the holes in the calico for the buttons. Pick or string
on more hair for the buttons, and finish as shown. The
circular box should be made with bottom to screw off: it can
then be lined. Tack or glue a strip of the covering round

FIG. 21.

the top edge of the box. The lining to finish is sewn to this
on the inside edge, and the covering on the outside. The
festoons are of course equally divided, four inches being
allowed for fullness across each. Draw them up on a thread,
and finish with cord and fringe, the shaped pieces imme-
diately under the cover or cut being separately in buckram,
finished with fringe and cord heading. This, being fixed to
the box, allows the cover to rise unencumbered and free.

Box Ottoman.— In Fig. 22 we show a square or oblong box ottoman to be upholstered in any two harmonizing colors, or in self-colors. The top of this should be of birch, or any hard wood, as soft wood will twist with the strain. Finish the top of this in the same way as described for circular ottoman (Fig. 21); but, the sides being square, the fullness will be the same each way in the buttoned border, that is, $1\frac{1}{2}$ inches. The festoons should be cut separately, tacked to the box, and finished with a cord heading. The bottom should screw off, so as to enable the workman to make a good job of the lining. When completed, the ottoman should not stand more than 18 inches high.

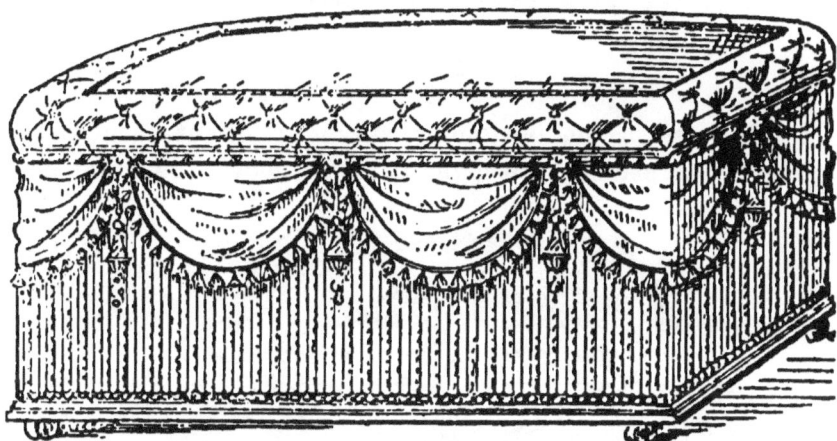

FIG. 22.

Double-pouffe Ottoman.— This ottoman, represented in Fig. 23, is made to look like two cushions placed one on the top of the other. These can be fixed or worked on a center swivel. The bottom cushion is composed of a wood frame covered, the border shown being fixed to the wood with nail buttons. The top cushion has a wood frame equal to half of the bottom one. The springs will be placed in this, and the stuffing rounded off to the shape of the bottom half.

The border will be buttoned according to instructions given
for circular ottoman (Fig. 21), and finished as shown in the
present illustration.   The ottoman being square, 1½ inches

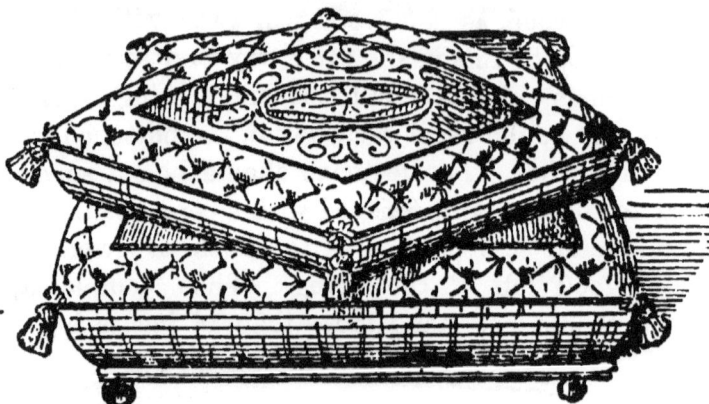

FIG. 23.

for fullness in the buttoning will be sufficient each way.
The rough sketch here given (Fig. 24) shows the frame as

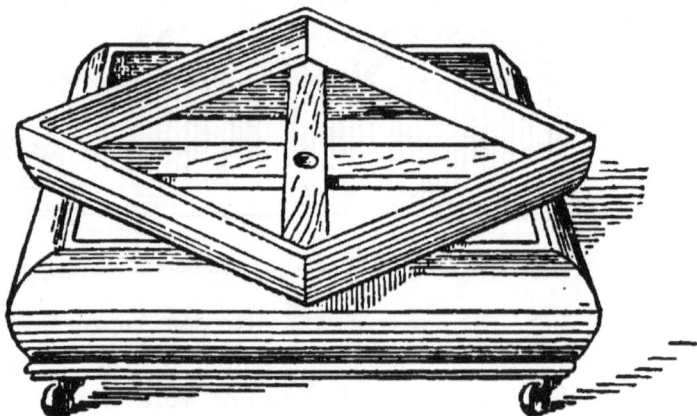

FIG. 24.

it would reach the hands of the upholsterer.   An ottoman
like this is very suitable for needlework, with borders of
plush to harmonize.

Gipsy Tables and Mantel-boards.—Gipsy tables, mantel-boards, &c., are often bordered with needlework. They should be shaped up on buckram, and the back lined with a material of a color similar to that of the groundwork, the lower edge being finished with fringe, according in its general character with the needlework. The upper edge of the border will be sewn to the upper front edge of the mantel-board or table-cover, and finished with a cord. This will cover the stitches, and should harmonize with the fringe below. Borders of needlework should not be strained, but allowed to go as free as possible. There is always a difficulty in getting borders to fit nicely where the article is square-cornered or of octagon shape; but, by having a light wood bracket of the depth of the border fixed under the table-top, the work can be made to appear as sharp and true as possible.

CHAPTER VII.

BEDROOM FURNITURE.

Spring Mattresses.—There are many patterns of spring mattresses now in the market. I will, however, confine my remarks to the old-fashioned box-frame spring mattress. In its construction there would be required eight laths across the bottom, and the sides should be about six inches high. A 4-feet 6-inch mattress will require forty 10-inch springs, that is, five for each lath. These will be fastened to the laths with small staples, and then tied down so as to make them a little rounding, after which they must be thoroughly lashed each way.

Strong canvas should be used for covering the springs, and these must be sewn firmly to the canvas, as previously described in the chapter devoted to the stuffing of a single chair. A well-stitched roll must be fixed round the box, and it should be about three to four inches high. Pick on the canvas about 20 lbs. of hair or 25 lbs. of wool; place the tick over and temporarily tack it; let the stripe of the tick run from head to foot; then tuft the tick and turn the whole upside down, tacking the tick on the lower edge of the box. Nail double webbing on the under side, and about one foot from the corners, for handles. Cover the under side with canvas, and the mattress is then finished.

If a mattress is required with a tufted top and welted or bound border, an extra three quarters of an inch to the foot should be allowed each way for fullness, when cutting the tick for the top. Mark out the top for tufts about six inches from the edges, making the diamonds about twelve inches to fourteen inches. Cut the border to the exact size required, and rather tight all round. The top will be made to come in the border by a small plait opposite each outside tuft. If the tick is a red or blue striped one, pipe the border with red or blue. It looks more suitable than binding on a spring mattress. The top will then be quite free and soft, and will present a good appearance.

For the convenience of removal, spring mattresses are frequently made in two halves. In such a case, the sides of the spring-boxes will each be half the length of the bed. There will not be any ends to go across the middle of the bed. The springs will be fixed as previously described, except that the two middle rows of springs should be so arranged that they may nearly meet. A piece of cane should be lashed across the two ends of the half-boxes where they join in the middle of the bed, so as to keep the boxes square, and at the same time form a base to work on. This cane must be stitched up all round, keeping the middle soft, otherwise the cane will be felt when the bed is used. Border and finish each half as before directed, keeping both the same height, so that they may correspond with each other exactly when placed on the bedstead.

Folding Mattress.—Another kind of spring mattress is made so as to fold up in two halves, like a book. The top of such a mattress will be all in one piece, the boxes being similar to those previously mentioned, but only about five inches high. Put the two half-boxes together on the bench, fixing the springs as if the mattress were in one piece. Lash the springs each way, and take care to keep the top flat, otherwise the mattress will not fold easily. Place the canvas over the top in one piece, cutting it a little at each side where the fold occurs. Cut the tick also in one piece, allowing three quarters of an inch to the foot for fullness. Cut a border 4½ inches wide, and so as to fit moderately tight, allowing two inches for turnings in the center fold. Having marked the top, case it into this, as previously described. A second border must now be sewn to this, to make it sufficiently wide to tack in under the bottom of the spring-box. Run a twine round on the edge of the case, and pick on a good body of hair. This should be very firm, as there will be no stitched up roll. Place the tick over, and tack it temporarily in position. Tuft the mattress, and stitch the border round, as before described for an ordinary mattress. The canvas and tick are the only connection that now holds the two halves together. Fold one half on the top of the other; cover the open ends with canvas or tick, sewing it to the cut borders; cover two pieces of web with the same kind of tick, stitching a piece of cane or spring-wire, about three inches long, crossways to the end of one, with a button large enough to take it in the other; nail these to the bottom of the frames, one on each half, so as to secure them together when folded for removal. This is an expeditious method of making a spring mattress that will stand a considerable amount of wear.

If the tick be of check pattern the appearance will be much improved by cutting the top border on the cross, and the bottom one on the straight.

If a spring mattress is made in two parts, it is, however, better that, instead of being in two *equal* parts, the foot part be made only one third of the whole length, and the head part two thirds, so that the join shall not come in the center, or where the greatest weight of the body would be.

If any of these kinds of mattresses are to be made with a spring-edge, the wood sides will not be required, but merely a lathed frame as a base for the springs. The latter will be fixed standing flush to the edge of the frame; then lash them well to shape. A piece of stout cane must be lashed to the top edge of the springs all round the outside to preserve the form. A roll will then be worked round to this, and it will be finished in the same way as the mattress previously described.

Hair and Wool Mattresses.—In cutting the ticks for these, measure the bedstead and allow the tick three quarters of an inch to the foot larger for fullness. The tufting will absorb this, leaving the mattress, when finished, of the correct size. The borders are usually cut 4½ inches deep, not allowing anything for the binding. A good-sized mattress cut in this way would be fairly filled with 9 lbs. of medium hair or wool to each foot across the mattress. A 3-foot mattress would require about 30 lbs., this being the smallest quantity that can be fairly used. Tuft the mattress with a 10-inch or 12-inch diamond, and about six inches from the edge. The better the quality of the hair or wool used, the better will it fill out in stuffing. A French mattress or pallet would take about 6 lbs. to the foot. French mattresses being made of half wool and half hair, the ticks are cut without a border. They will require an allowance of 1½ inches to the foot each way for fullness to make them finish the right size. Only one side has to be sewn together, as there is no border. Spread one half of the cover on the boards, and let the other hang down out of the way. Let us suppose that 18 lbs. of wool and an equal quantity of hair are required in a 5-foot mattress. First, put 9 lbs. of wool evenly over the half tick, then the 18 lbs. of hair in the center, and the remaining 9 lbs. of wool on the top. Pull over the other half of the tick, pin it round, and stitch it. It now only requires tufting, and the job will then be finished, making a soft and comfortable mattress suitable for using on the top of a spring bed. If this kind of mattress is to be stuffed with all hair, it should be sewn round before filling it. The only object of having the mattress open is to keep the wool and hair separate.

Bedding.—A good feather-bed tick should be cut with a 5-inch border, and be of the actual size of the bedstead, simply allowing for turnings. The stripes should run lengthwise of the bed and crosswise on the borders. The welting will also be formed of the tick material, cut on the cross. About 8 lbs. of feathers to the foot in the width, and these of medium quality, are required to fill a bed. For instance, for a 5-foot bed will be required 40 lbs. of feathers. Of course, the better the quality of feathers, the better they will fill. Bolsters are cut 20 inches wide (40 inches round), and their length is determined by the width of the bed. Bedding manufacturers usually cut bolsters and pillows 36 inches for stock sizes. The ends are gathered and welted into an oval piece about 12 inches by 8 inches, or to square piece 14 inches by 6 inches, rounded on the ends thus (+   +). Half circle with the compass from + mark, the welt being cut on the cross. The two pillows are also cut 20 inches wide, and, of course, half the length of the bolster: they are simply sewn to finish square. Bolsters require 7 lbs. of feathers, and pillows 3 lbs. each. These weights, if for 5-foot bedding, would only be sufficient if of *very* good quality.

Common flock-beds are usually cut to the size of the bed, allowing for turnings, but without a border. The bolsters and pillows for flock-beds are finished square, and the quantities for filling are about the same as those given for common feathers.

CHAPTER VIII.

BED DRAPERIES.

To take the correct measure for bed draperies on an iron or brass bedstead, the bedstead must first of all be put up. Measure the length and depth of base-valances for the sides and foot, and the drop for curtains from the curtain-rod to the floor. Take the height and width of head, allowing sufficient in width to pass round the head pillars and to tie with tapes from side to side at the back. If the bedstead be of wood, the material can be fixed with tacks. Measure for size and shape of tester-piece, and for the length round for the tester-valance. Generally there is a wood frame made for the tester in iron bedsteads, the same shape but about half an inch larger all round than the iron-work. The cost of this frame is trifling, and it keeps the tester material from bagging down. It also enables the workman to fix the valance firmly, so that it can not be moved out of place, as is frequently the case with those which are simply tied to the ironwork with tape. The head and tester-cloth should be backed with forfar or strong cotton. In cases where there is no wood frame, the best way is to cut the tester-cloth nicely to shape. Sew the valance to it before hanging. Fix to the ironwork with tapes sewn about six inches apart on the top inside edge of the valance.

When cutting the base-valances, let them reach to within one inch of the floor, finishing them with a hem at the bottom. In order to fix them to the bedstead, have some wood laths made about two inches wide. These will rest flat on the iron, keeping the valance rigidly in place. Cut strips of material wide enough to go round the laths, and finish them

with a binding to the top edge of the valance. A few tapes
will hold the laths firmly on the bedstocks.

The curtains when finished should just touch the floor.
The top and bottom turnings will require about four inches.
If the tester-valance be of dimity, it will look well with a
box plait,— that is, a 3-inch plait and a 3-inch space, and
about 16 inches or 14 inches deep; or it could be cut about
12 inches deep in the center, and sloped to 18 inches at the
ends. Finish the lower edge of the valance with a fringe,
and the upper edge with a rope-cord of about 2 inches or
1½ inches diameter, made for the purpose and matching
with the valance; or make a frilling of the same material,
1-inch plait and 1-inch space, about 3 inches wide. This
can be sewn or tacked on. This will also require to be cut
twice the length required to finish, and gives quite an effective appearance. The valance must be cut twice the actual
length which it is to be when finished,— the extra length
will be taken up in the plaits. The same remark applies to
the base-valance, which should in all cases be finished with
box-plaits. Should the bed furniture be of tapestry, or any
other material under which a plain buckran can be placed,
it would be advisable to do so, but a straight round valance
always looks well, and so does a graduated one, increasing
from 12 inches in the center to 18 inches at the sides. We
append sketches of half a dozen different valance shapes,
which can be used either for window or bed draperies, since
they can be lengthened or contracted in the straight parts,
as may be necessary. The measurements indicated will be
found, if followed, to yield graceful and well-proportioned
valances. Instructions for setting out the valances will be
given later on.

Persian Beds.— It is better to have a wood frame made
for these, and to canvas and button it as illustrated by
sketch (Fig. 25). The stuffing may be done very soft, and

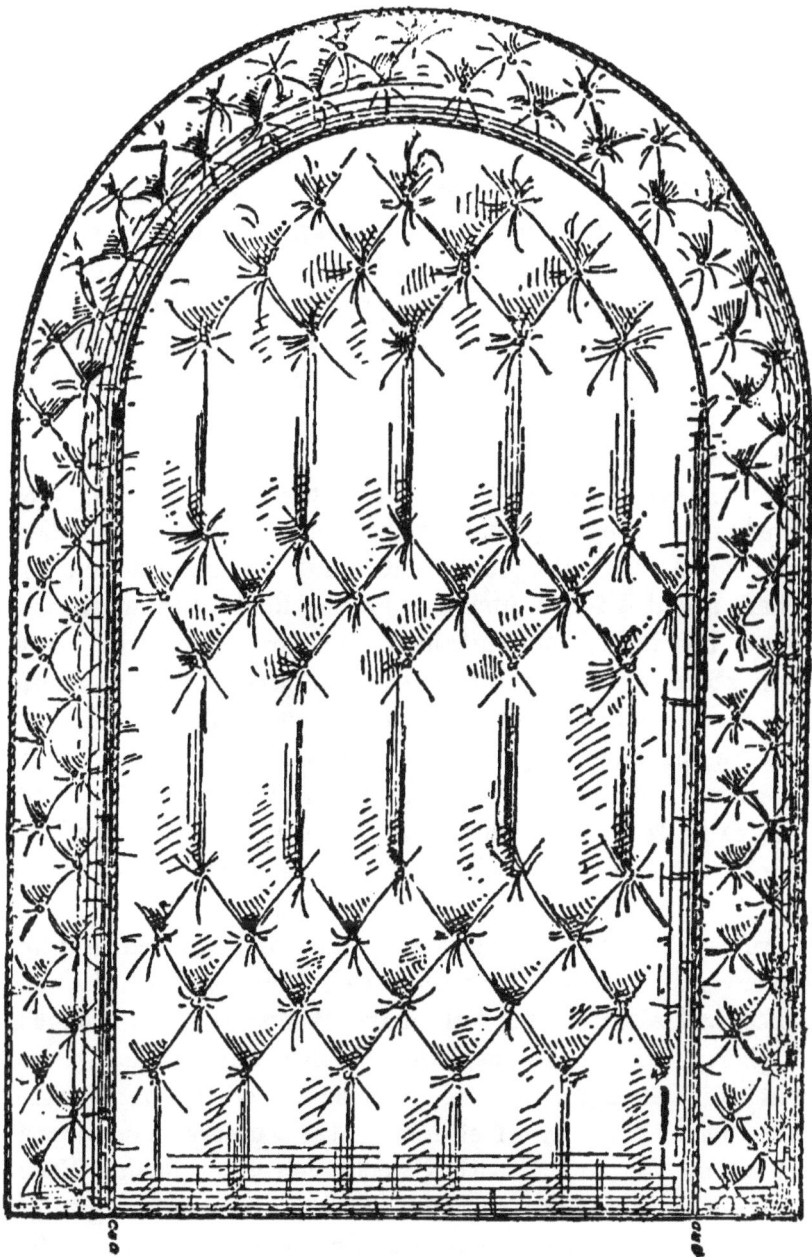

FIG. 25.

the covering should, if possible, be in two colors, the outside roll or buttoned margin in one, and the inside in another, using a bold cord to cover the joint. A few tapes tacked to the wood frame will be sufficient for tying it firmly to the ironwork. If the bed be of dimity, cretonne, or cotton material, allow 1½ inches for fullness; if of rep or tapestry, 1¾ inches; but in the length of the flutes no allowance will be required. The base-valances are cut, as before directed, a 3-inch box-plait and a 3-inch space alternating with each other. Double the length of valance material will be required to finish to the proper length.

Window Draperies.—Cut the curtains so that when they are finished they will touch the floor, allowing about 3 inches additional. Fix the hook that holds the curtain-band or chain about 4 feet 6 inches from the floor; and when the chain has pulled the curtains to their proper position they will about touch the carpet. At the present time valances are mostly cut straight, or with a slight break in the line to relieve the monotony. The depth will vary, and must be governed by the height of the room: they range from 14 inches to 27 inches. Should there be a mass of wall-space above the top of the window-frame, as is often the case in old houses, matters may be greatly improved by fixing the cornice-lath above the window on the wall-space, calculating to cover as much of the wall-space as the depth of the valance will allow. This will make the window look much higher. Valances for bay-windows have a good effect when cut quite straight. If the room be moderately high, make the valance about two feet deep, and if tapestry be employed it should be paneled with Utrecht velvet or plush. Each window might have a panel, leaving a 6-inch margin of tapestry round every panel, and finishing the edges of the velvet with a gimp. The base of the curtains should also have a width of the same velvet sewn on across, and about a foot from the ground. The velvet is 24 inches wide, and when thus applied it answers the purpose of a dado. Care must be taken in the selection of the colors, so that a harmonious effect will be produced.

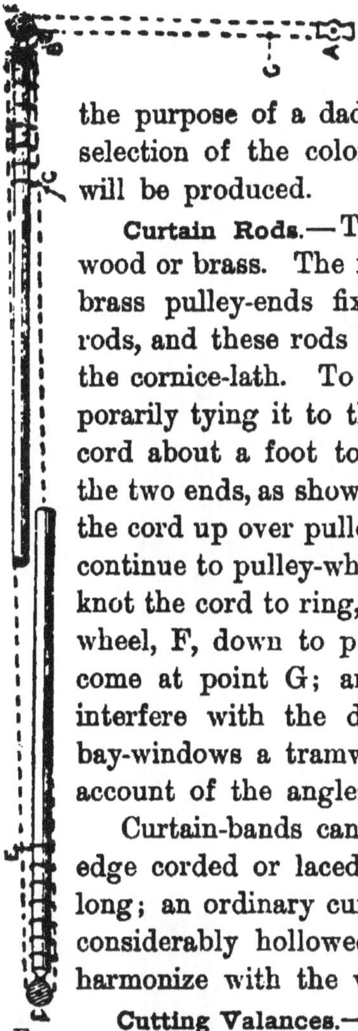

Curtain Rods.—These are made of one-inch hard wood or brass. The former are usually of birch, with brass pulley-ends fixed at either extremity of the rods, and these rods are fixed with drop-hooks under the cornice-lath. To fix the cord, commence by temporarily tying it to the pulley marked A, leaving the cord about a foot too long; pull the rings back to the two ends, as shown in sketch (Fig. 26). Now run the cord up over pulley-wheel, B, and knot to ring, C; continue to pulley-wheel, D, passing under to the top, knot the cord to ring, E; continue over top of pulley-wheel, F, down to pulley, A. Then tie the knot to come at point G; and in that position it will not interfere with the drawing of the curtains. For bay-windows a tramway action will be necessary, on account of the angles.

Curtain-bands can be made of same material, the edge corded or laced; average run about 18 inches long; an ordinary curtain-ring each end, the top end considerably hollowed from the middle, shaped to harmonize with the valance.

Cutting Valances.—The accompanying illustrations of valances (Figs. 27 to 36) will be found adapted to represent everyday requirements, and they are easy to cut. One half of the sketches show the patterns, and the other half the measurements. They are all alike suitable for window or bed draperies; and having cut them many times,

Fig 26.

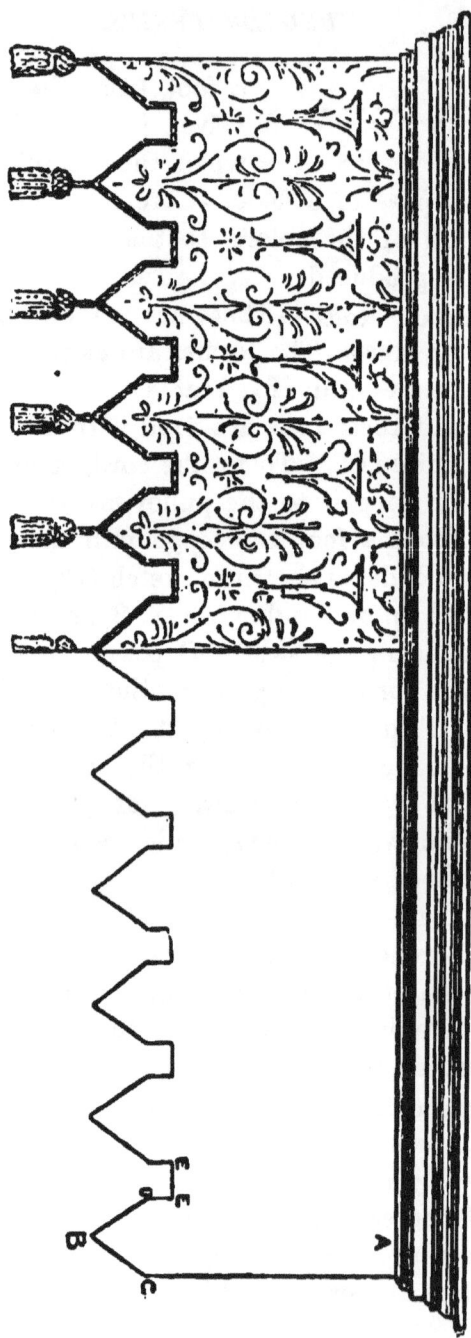

FIG. 27.

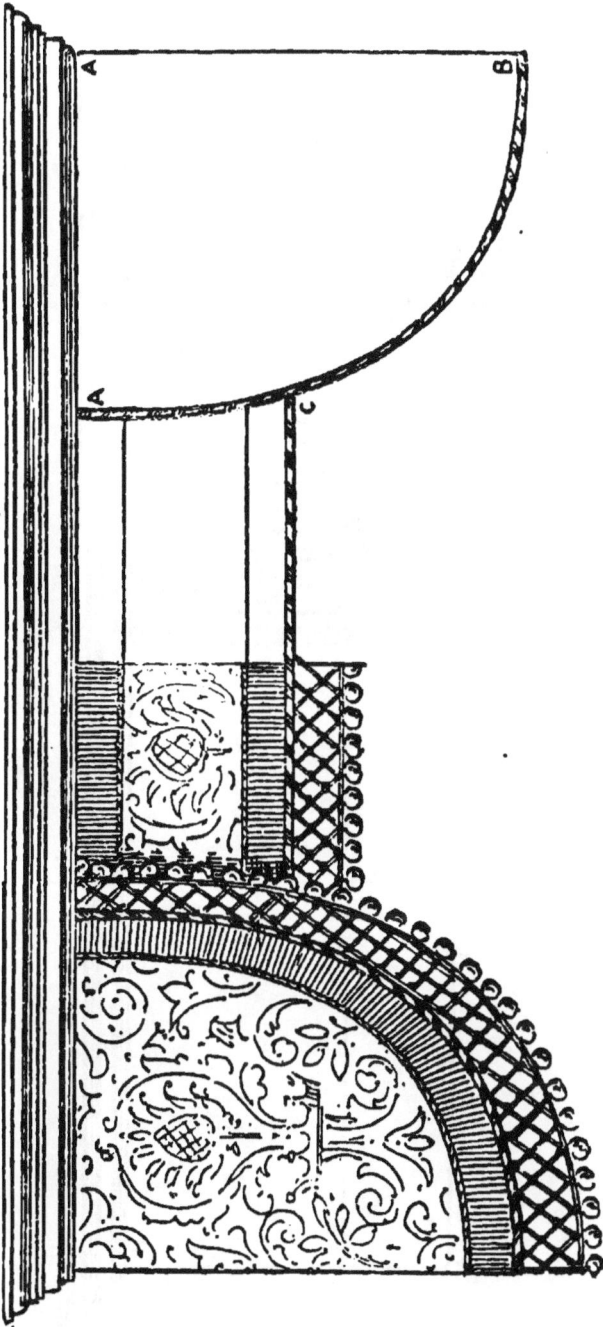

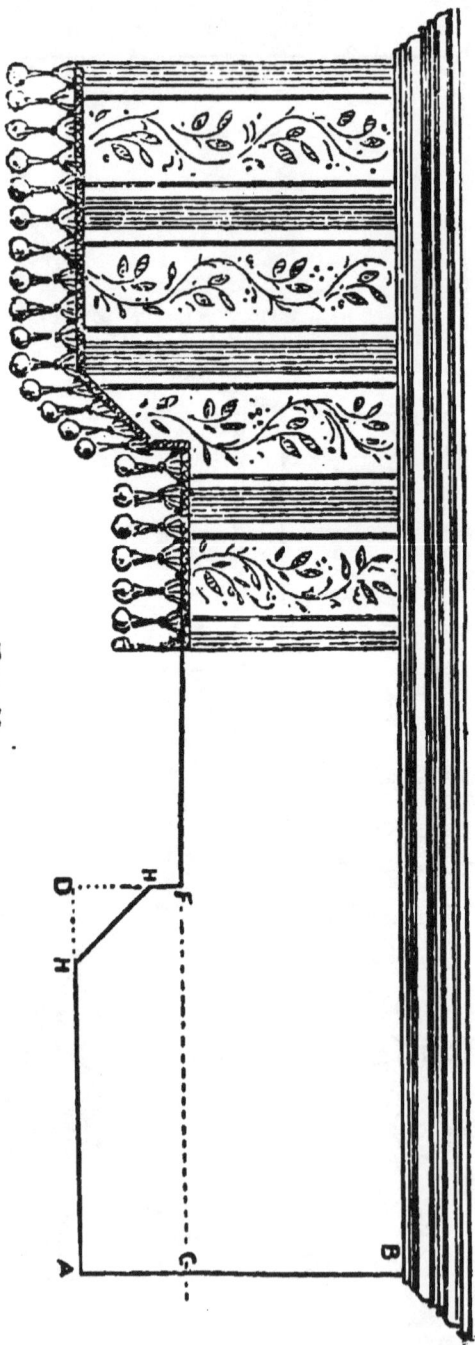

FIG. 29.

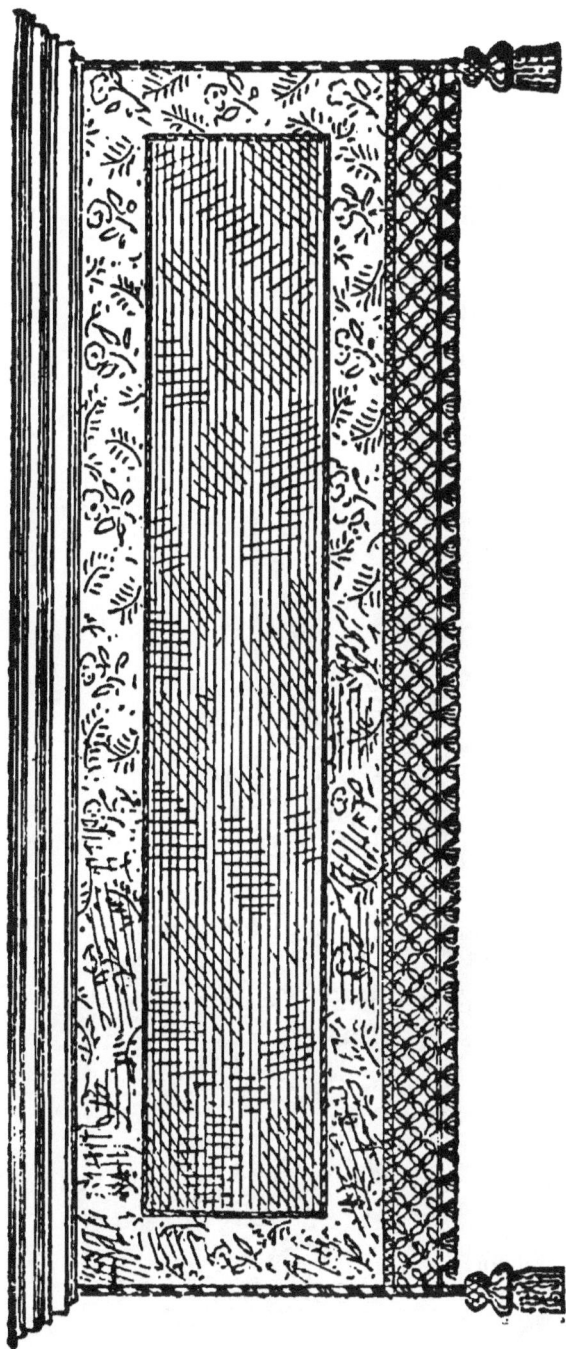

Fig. 30.

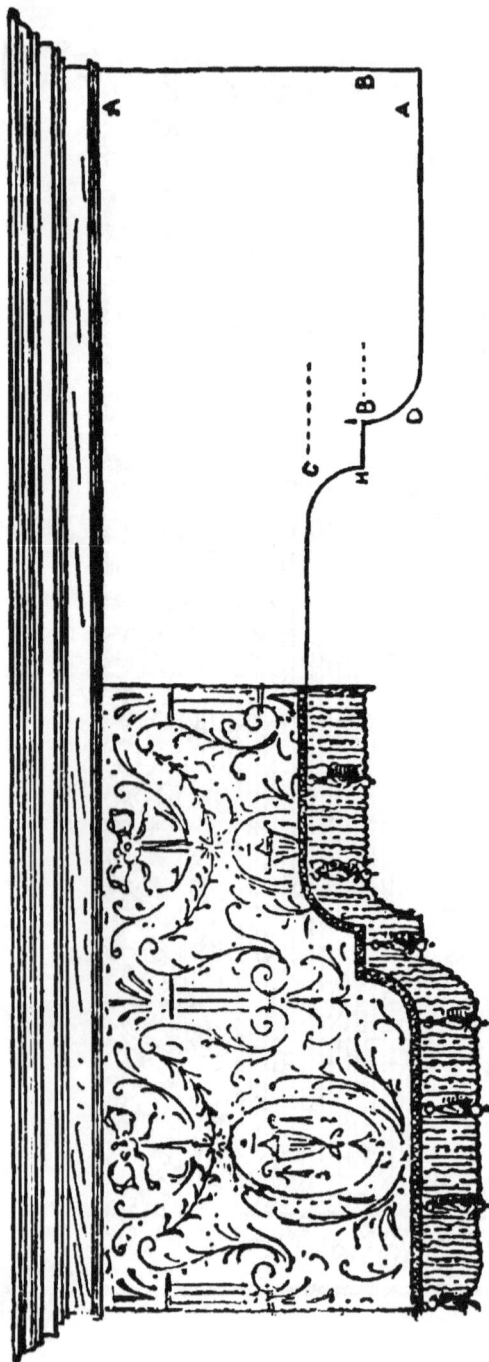

Fig. 31

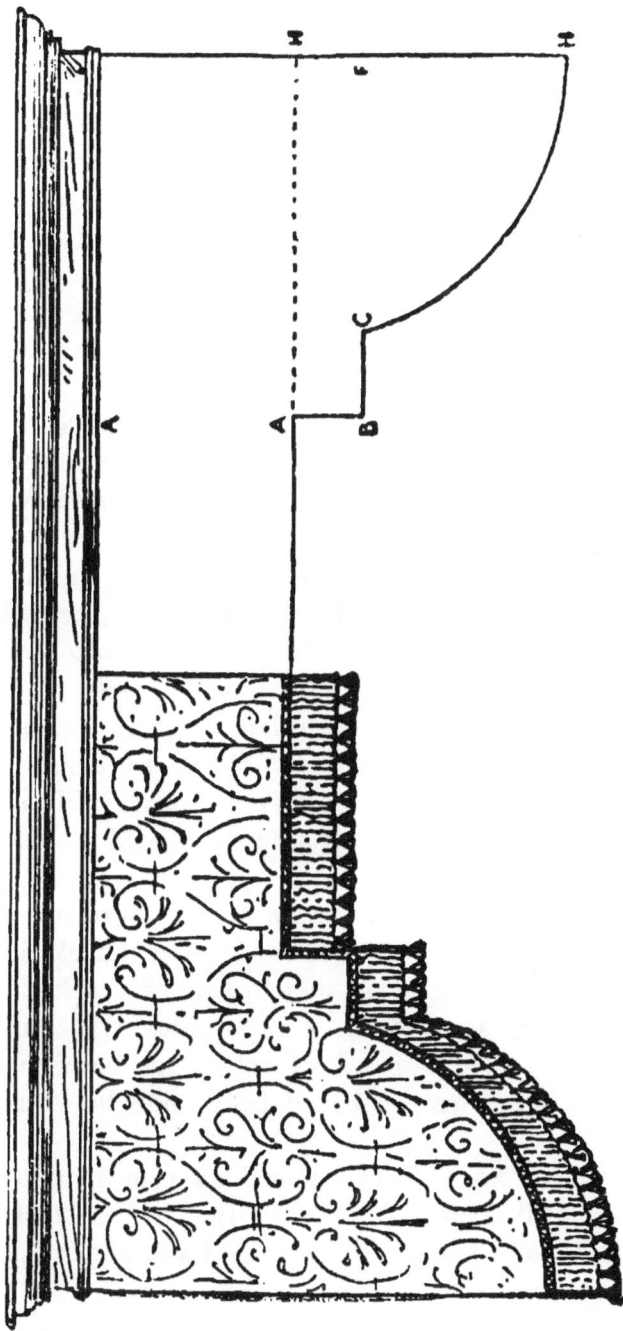

FIG. 32.

the writer knows from practical experience that they look well when completed.

No. 27 valance is cut 16 inches deep from A to B; from C to D, 4 inches across; from E to E, 2 inches; from E to D, 1½ inches; from B to the line shown at E E, 4 inches. Finish with a small cord all round the edge of the pattern as shown, and with tassels at the extreme points.

The valance shown in Fig. 28 is in one piece of buckram, the measurements being as follows:—A to C, 12 inches; A to A, 20 inches; from A to B, 24 inches; plush border, 3 inches, and rounded as shown. To vary this, you may have a straight buckram for the center, same as shown, and gather or plait up the wings separately. In this case, the depth of the wings would be the same, but the width would necessarily have to be doubled to supply the fullness. Should the window be a large one, the depth of the whole should be considerably increased, say from 4 to 6 inches.

In Fig. 29 we illustrate a valance cut 18 inches deep from A to B; the distance from A to C is 6 inches; from C to F, 20 inches; from D to H, each way, 4 inches. Cut off the corner, which will be at 45 degrees, and the shape as shown will be obtained.

No. 30 valance is quite straight, and is paneled with velvet. It should be 18 inches deep. The velvet panel is 10 inches wide, leaving a tapestry margin of 4 inches. This pattern is very suitable for a bay-window. If thus employed, arrange for a panel of velvet for each window, showing an equal margin above, below, and between each panel.

The valance No. 31 is 18 inches deep from A to A; from B to B it is 20 inches; from H to B, 3 inches; and from lines C, B, D, each 3 inches. Strike from a radius of 3 inches for both the curves or sweeps.

Another of the valances shown (Fig. 32) is 12 inches from A to A; from A to B and from B to C, each 3 inches; from H to H, 12 inches; and from C to F, 18 inches. Strike from an 18-inch radius for the sweeps. The downward curve in the straight part will have its center at a point bisecting the top line. Some other suggestions are shown at Figs. 33, 34, 35, and 36. Fig. 33 would form a novel and attractive piece of enrichment. Its disposition affords ample scope for an effective and artistic display of light and shade in drapery. The central valance is intended to represent a piece of figure tapestry, cupid gambols, or anything of a suitable character. This should be framed up with a plush border, and fringed at the lower edge. Upon either side of this, festoon drapery should be arranged after the manner indicated in the illustration. The character of the woodwork will admit of the introduction of one or more decorative items, such as a Japanese fan, tray, or plaque, which would considerably enhance the general effect. The lambrequin, composed of lozenge-shaped panels (Fig. 35) should be treated in a similar manner to that above described. The center of each lozenge should consist of a piece of tapestry, also inclosed by a plush border. The openings which occur between the panels admit of the visitor observing the beauties of stained glass, now so generally employed in the decoration of the upper part of the window-sashes. Pelmets frequently quite prevent the beauties of such enrichment being seen. The two remaining draperies, being simple in character, explain themselves. Those lambrequins attached by studs or buttons to the cornices are intended to be detachable, for the sake of cleanliness as well as for effect.

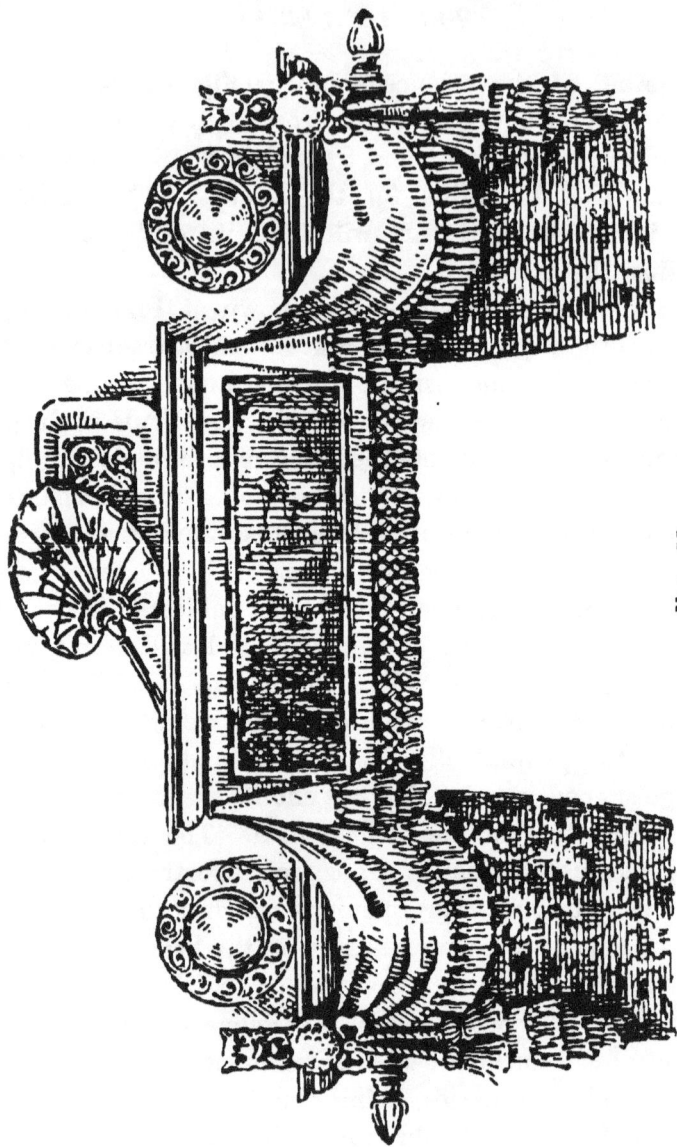

FIG. 33.

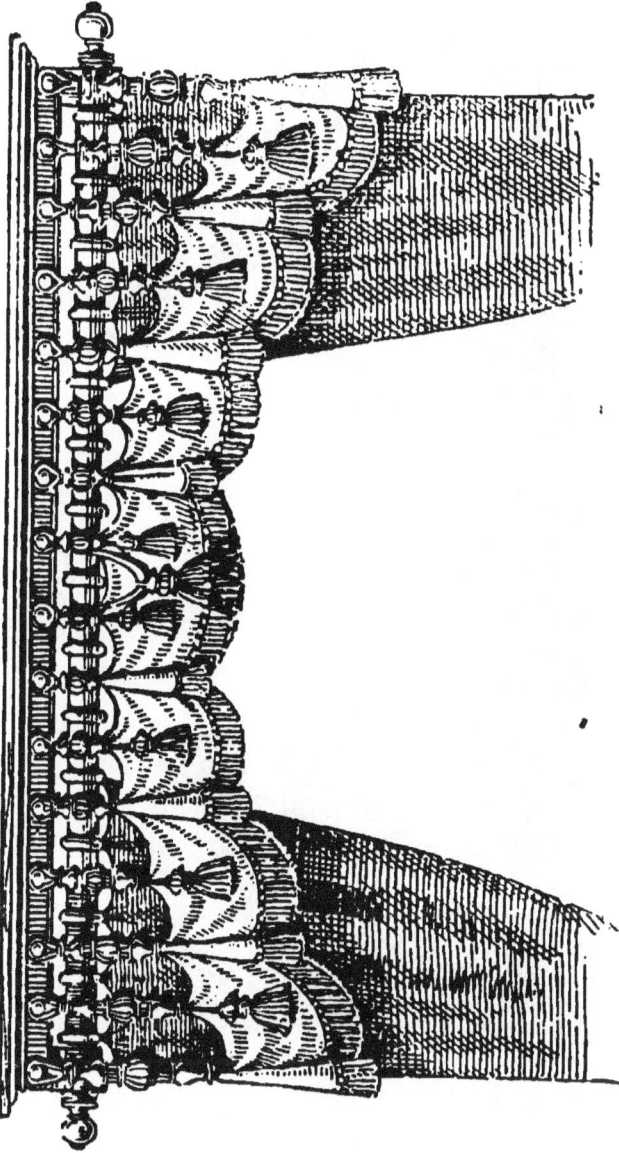

FIG. 34.

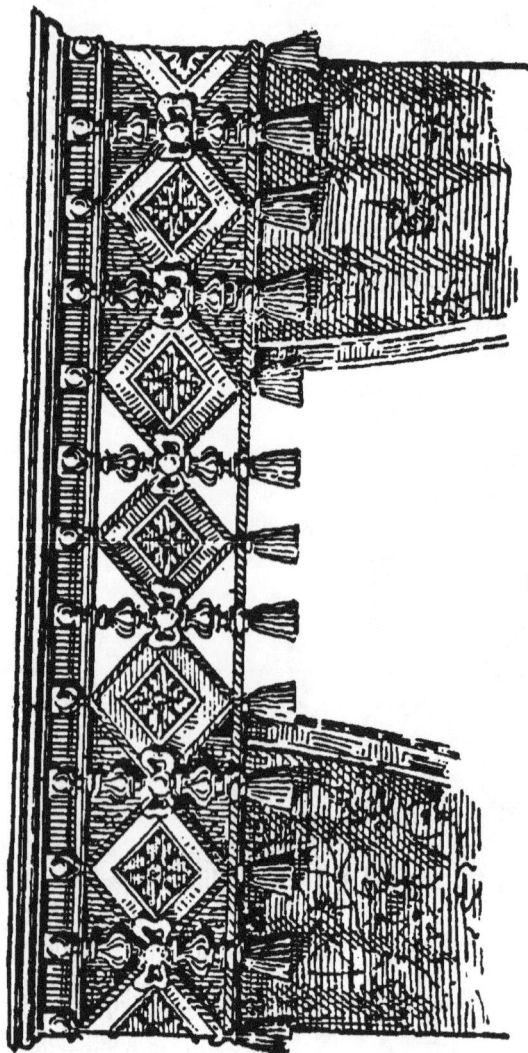

FIG. 35.

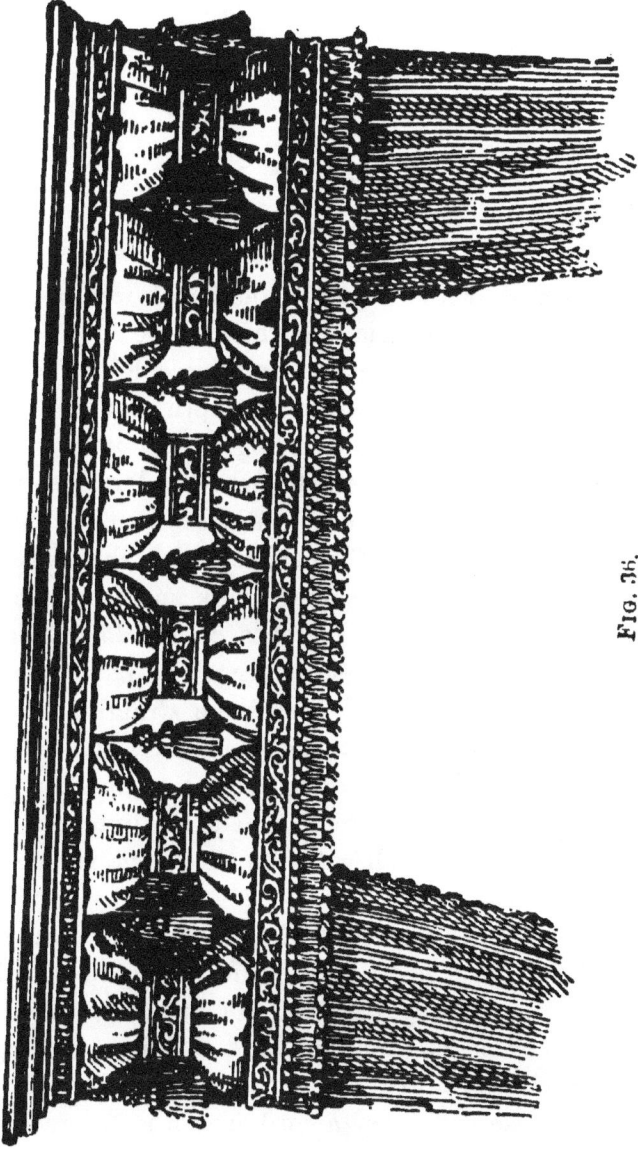

Fig. 36.

CHAPTER IX.

BED-HANGINGS.

THE design shown in Fig. 37 would look well in tapestry, or in any nice material in which there was not much stiffness. Let it be lined with a suitable color. The valance is cut in two pieces, and double the depth required to finish. This will be taken up in the reefing. The bottom edge in the center must be cut 5 or 6 inches wider than the top edge. The small center-piece is separate, cut in buckram, and lined in the shape of a bell, to finish about $3\frac{1}{2}$ inches wide. The two front outside corners should finish about 15 inches, graduating to 22 inches on the back edge, not including the tassels. These latter would be all the better loaded with lead, which will keep them straight. The head and tester cloths are made of the same material as the lining for the tapestry, or whatever it may be, plaited up on the face of it, as the sketch shows; and if the lining be a good shade and harmonize well with the tapestry a very pleasing effect will be produced.

Door or Window Drapery.—The drapery shown in Fig. 38 is to be fixed on a wood lath. Cut the actual width and depth required, simply allowing for turnings. A narrow tape should be stitched on the back of this at an angle of 45 degrees with half-inch rings sewn on at equi-distances of 3 inches. These will carry a cord from the fringed side of the curtain to a pulley-wheel placed in the top right-hand corner, and by drawing this the effect shown in the sketch will be obtained. An ordinary curtain-holder or chain completes the whole. The rosettes in each corner are of the same material. The wing or tails in the left corner would

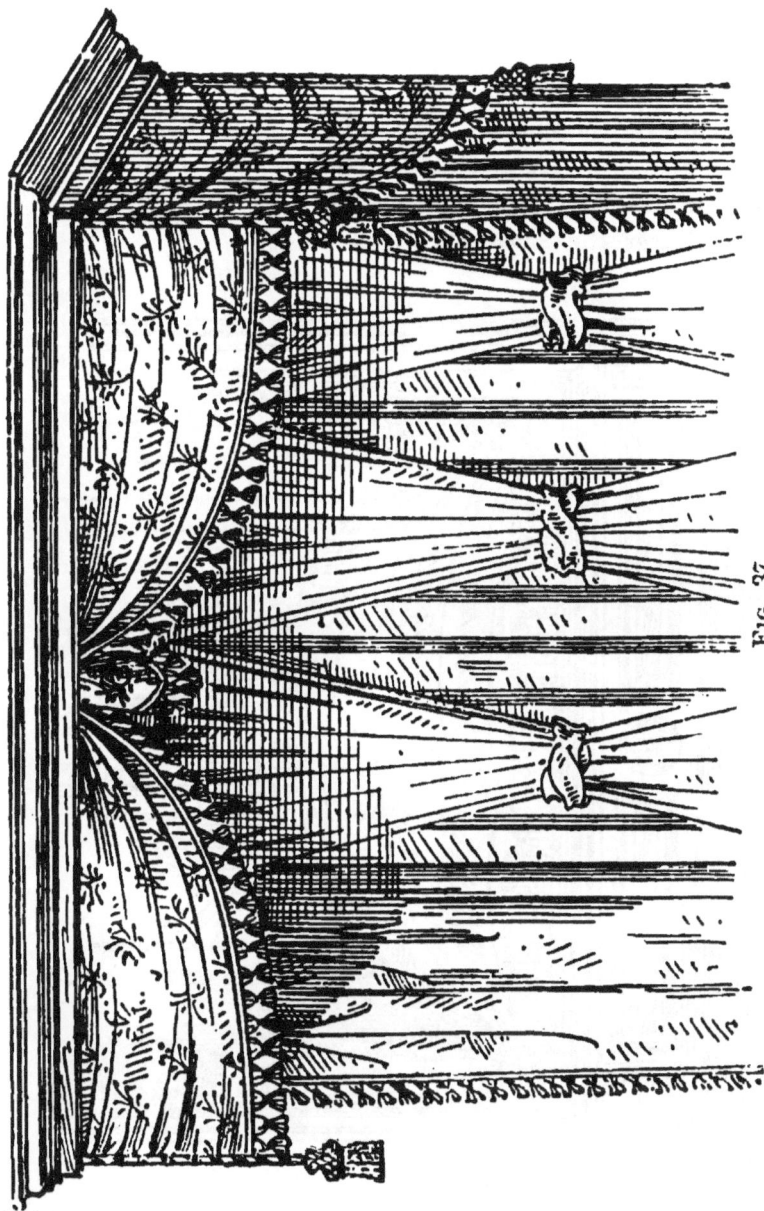

Fig. 37.

be cut about 20 inches wide, 22 inches deep outside edge.
12 inches on the inside. A straight line cut across and then
folded up will give the shape shown. For a doorway the
wing would be cut much smaller than for a window, for
which latter the sizes are given.

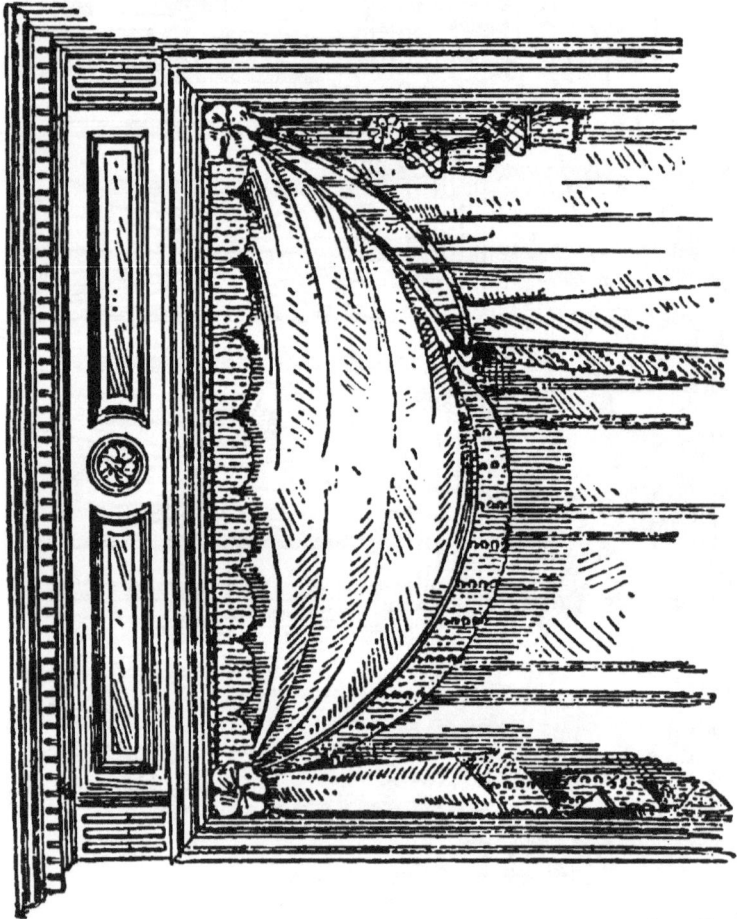

FIG. 38.

Mantel-board and Drapery.—To carry out the design
illustrated in Fig. 39, get a one-inch pine-board for the top
of the length required, and sufficiently wide to admit of the
rods upon which the curtains are to run being fixed thereto.

A drapery of this kind would look well in velvet, either embossed or plain. The board is covered and tacked in under, and studded on the chamfered edge. The festoons are cut

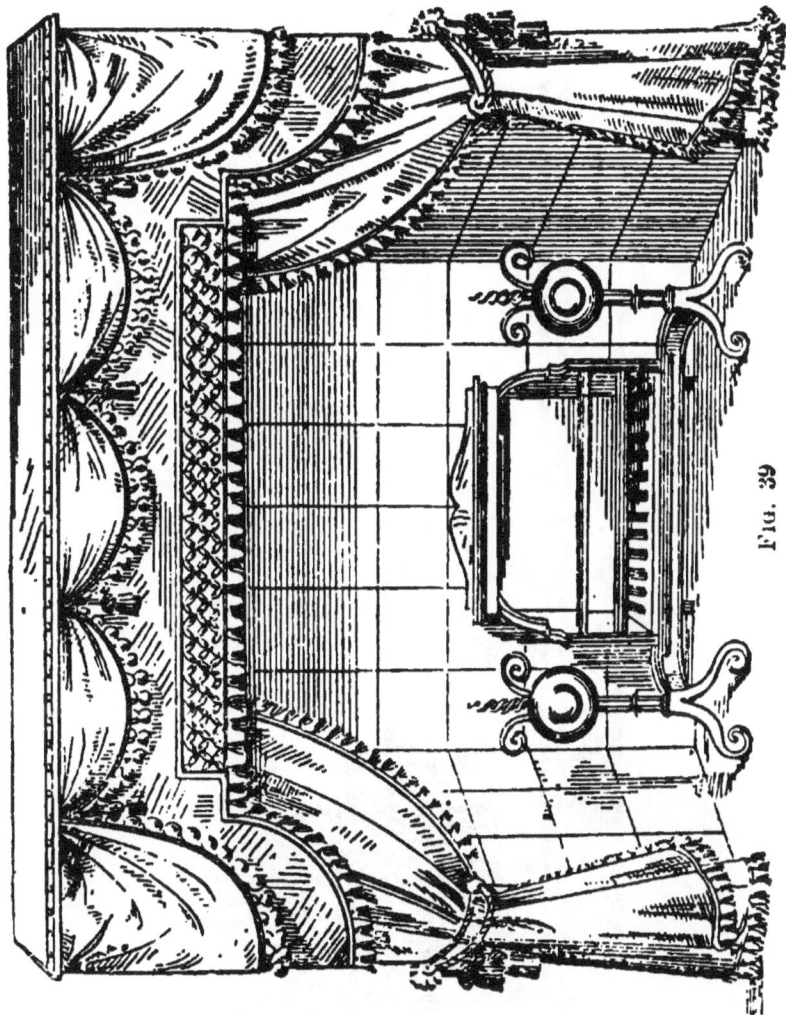

Fig. 39

10 inches wide on the top edge, the bottom edge being 4 inches wider and rounded; the depth is 12 inches. Each one is made separately, and fixed, as shown, to the strip of

wood under the board, as is also the lambrequin.   The lambrequin is made on buckram, according to this size, showing
half section.   The two corner festoons are cut considerably

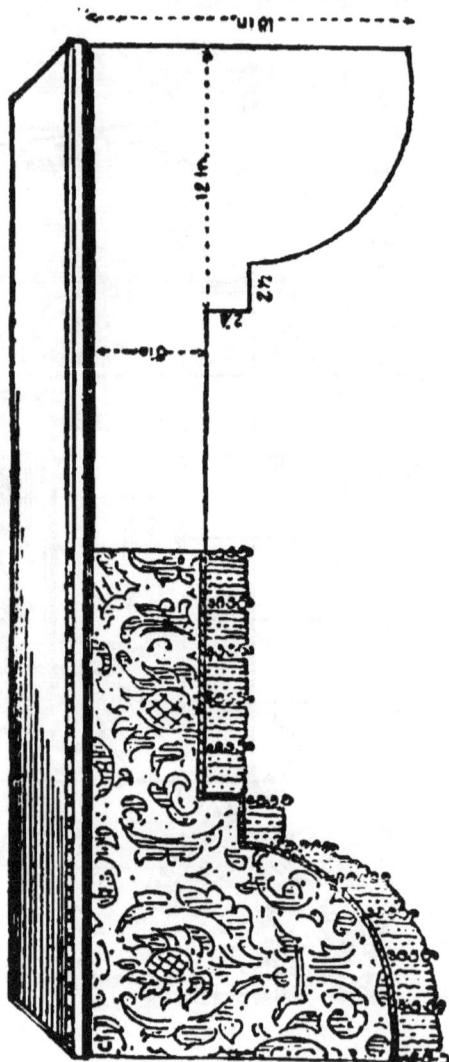

Fig. 40.

deeper than the others and returned on the ends straight,
as is also the lambrequin.   The above remarks apply to the
mantel-board shown at Fig. 40.

Figs. 41 and 42 show two other methods of draping a mantel. Fig. 41 looks extremely well in plush, with Persian trimmings in colors to suit. Fig. 42 shows a very elaborate mantel-draping that may be arranged to suit the color of the furniture, paper, or other decoration in the room.

Fig. 41.

Mirror Drapery.—To carry out the design shown in Fig. 43, the festoons are cut separate from the side-wings and heading-piece; the two center-pieces are also separate (covered pieces of buckram). The wings and tails are cut according to instructions previously given in connection with Fig. 38. Tack the whole to a strip of wood, which should be screwed to the top of the glass or wall.

FIG. 42.

Wicker-Chair Cushions.— For this design (Fig. 44) cut a paper pattern of the exact shape of seat and back, and cut the material in accordance with this pattern, simply allowing for turnings. When the cushions are filled and buttoned they will fit correctly.

FIG. 43.

By using a fairly good wool, a good cushion can be made without inside cases, and will answer every purpose. Wicker chairs can be greatly improved by stuffing the back and

sides on a piece of buckram. When this is stuffed, line the
back, cord the top edge, and fix it to the frame with a few
ties, directly under the plaited wicker-work. After this is
done, sew the fringe on so as to form a perfectly straight
line, as shown in the sketch.

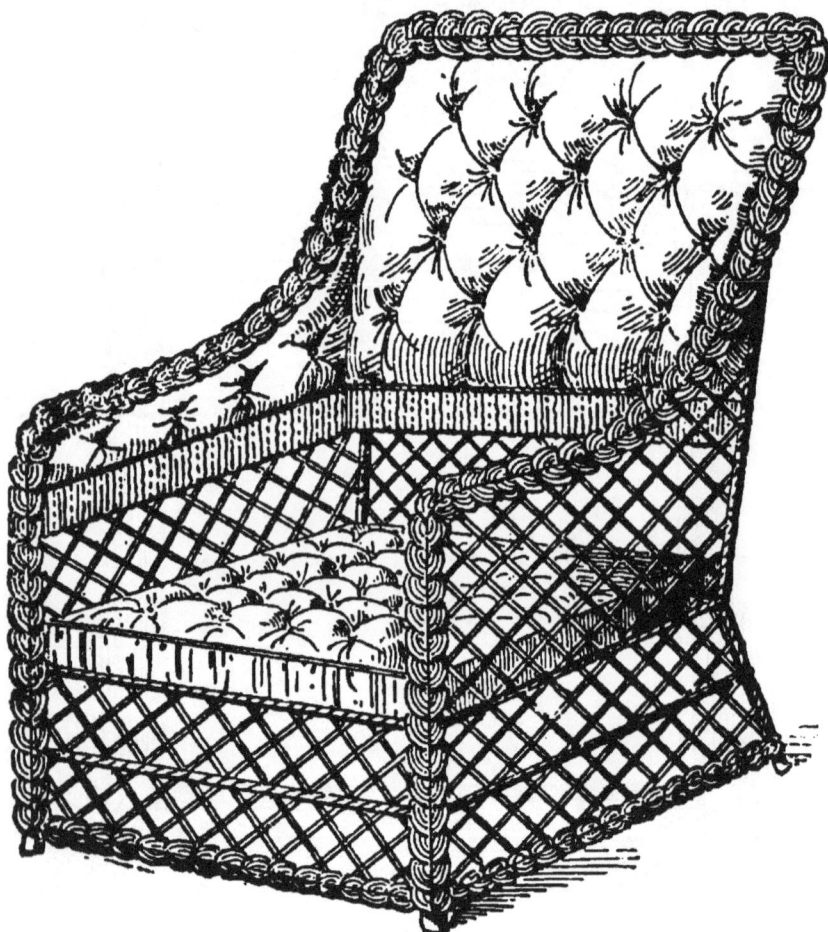

FIG. 44.

CHAPTER X.

CARPET-PLANNING.

THERE is nothing particularly difficult about carpet-planning, providing the measurements are carefully and correctly taken. A chalk-line should be always with you in carpet-planning to use in the place of a straight-edge, and is more portable. Supposing the plan of a room has to be taken. Let us first have a plan-book not less than a foot square: small sheets are awkward when there are many measurements to be taken. A 3-foot rule is the most convenient for the purpose of measuring, for the width of Kidderminster and Dutch carpet is 3 feet; and, moreover, by folding down one joint of the 3-foot rule we get a measurement of 2 feet 3 inches, the exact width of tapestry, Brussels, Wilton, and Axminster carpets. So that with a 3-foot rule, which always folds up to 9 inches, we can easily and quickly estimate how many yards of carpet will be required by simply multiplying the length by the number of widths.

Now as to the actual method of measuring. First take the square block of the room, and fill in all the details afterwards. Mark this down distinctly on a sheet of the plan-book, drawing as nearly as possible a miniature sketch or plan of the room, and sufficiently plain that it may be easily understood by any other person. It is well to bear in mind that the man who measures and sketches the room does not always plan and cut the carpet, hence great care is advisable in giving the proper figures. It is necessary to take cross-measurements, that is, from corner to corner, with a view to ascertain whether the room is out of square or not: there will then be no possibility of mistakes. The short measurements in a plan should also be checked by adding them together: if they correspond with the long measurements, the figures are correct. There is nothing like making sure that the plan is correct before leaving the house.

To cut the carpet, first lay down the principal measurements, or square block of the plan, on the cutting-floor by means of a chalk-line; and set it by the cross-measurements. Fill in all the other measurements afterwards. If these correspond with the figures entered for the square block, then all is correct. If one measurement on the floor is wrongly made, it will not come in correct at the finish; hence the absolute necessity of adding together all the small measurements of each side of the room, so as to test them with the long measurements before leaving the house. These additions are shown on the sides of the plan illustrated herewith (Fig. 45).

By examination of this illustration it will be seen that line C D is longer than A B by 6 inches, while the sides A C and B D are each 18 feet. It follows, therefore, that the room can not be square at some of its corners. The diagonal measurements, 23 ft. 5 in. and 23 ft. 9 in., determine this exactly. The larger angle is obviously one opposite the greater diagonal, namely, A B D. Also, as C D is a little longer than A B, it is likely that the angle opposite B C is the right-angled, or square, corner. In planning, this is easily found, but it can also be done by arithmetic. It is a rule in geometry that in a right-angled triangle if the two short sides be squared and added together they will equal the square of the long sides. In our example we have two sides, respectively 15 feet and 18 feet, and a diagonal 23 ft. 5 in. Squaring these, we have—

18 feet.	15 feet.
18 feet.	15 feet.
144	75
18	15
324 feet.	225 feet.

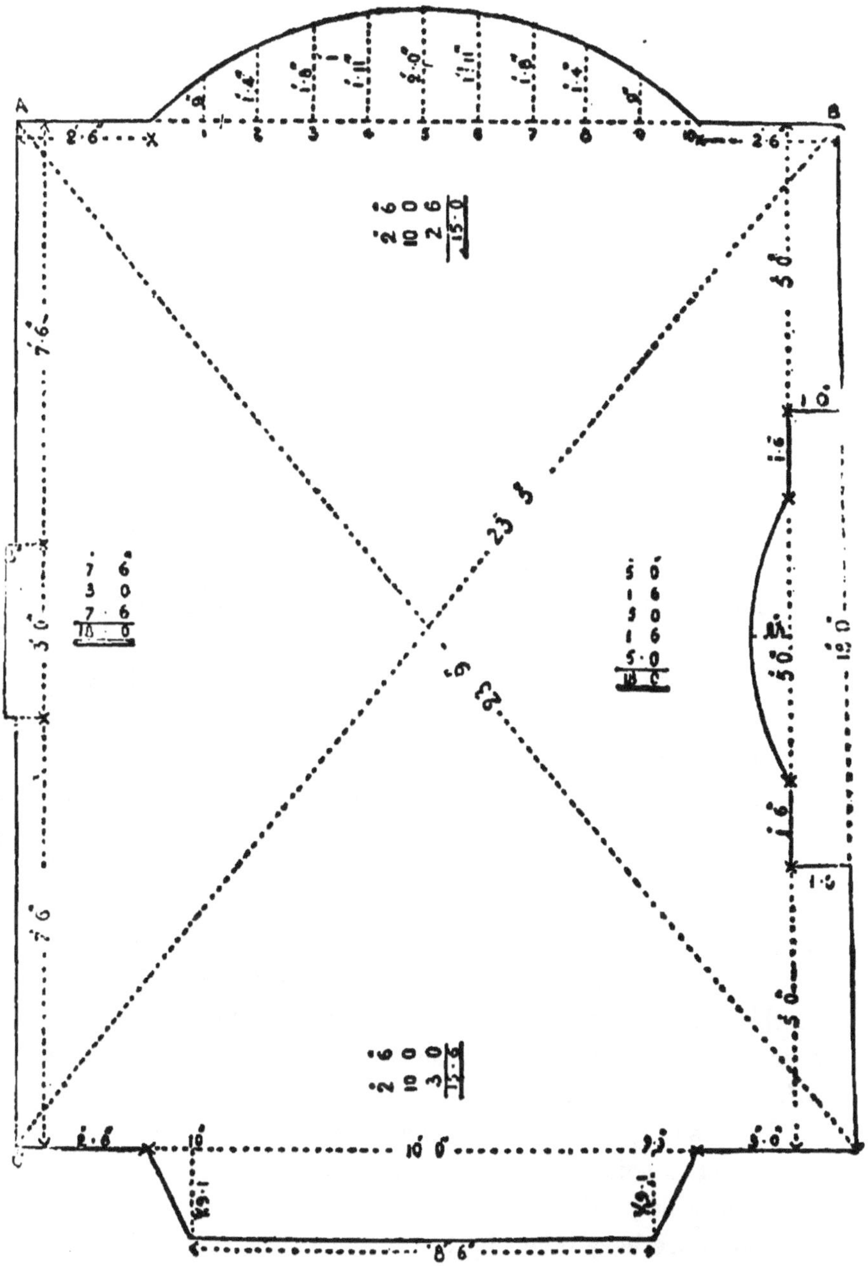

Fig. 45.

By cross-multiplication,—

$$
\begin{array}{rrr}
23 & 5 & 0 \\
23 & 5 & 0 \\
\hline
9 & 9 & 1 \\
538 & 7 & \\
\hline
\text{feet, } 548 & 4 & \overline{1} \\
\hline
\end{array}
$$

Adding 324 feet
to 225 feet
—
gives 549 feet.

This is so near the 548 feet 4 inches found by squaring the
diagonal as to show that the measurement of 23 feet 5 inches
is correct within a very small fraction of an inch, the angle
B A C being square.

On the same principle are founded the two following
variations of the above. A simple method of finding whether
the sides of a room are square with each other is to measure
along one side, as A B, 12 feet; and 12 feet along the other
side, A C. The diagonal, as B C, will then measure 17 feet
within a very minute fraction of an inch, if the angle, B A C,
is square. As a 5-foot measuring-rod is a very common size,
it may be still more easily used than the above, by measuring from A toward B 3 feet, and from A toward C 4 feet,
marking a point on the floor against the skirting in each
case. The distance between these two points will be exactly
5 feet, if the room be square.

Proceeding in a similar manner, these methods may be
employed in setting out the plan on the cutting-floor.

A circular window is also shown on the plan. The measurements are here taken by offsets. In the present example
these are ten in number, and one foot apart, because the
span of the window-opening is ten feet across. Mark these

offsets on the floor, in a direct line with the straight wall. Measure the exact length of each offset and mark it plainly in the book, as shown. The segment can then be easily laid on the cutting-floor.

When cutting bordered carpets the border should first be cut, taking care to have a good miter at each angle. The body is next filled in.

Seams across the width of a carpet, called "cross-joins," should have no turnings; and if sewn closely will stand any strain and not fray out. Joins running the selvage way, or where it is necessary to put a part of a width, should be cut about one inch wider than required to finish, and frayed out, and can then be sewn as if it were a selvage edge. Sew it close, and it will stand any strain. By this method you have an equal thickness, and prevent the joins wearing before the other parts. Miters, of course, you are bound to turn in.

Felt carpeting is rarely planned to fit. It is, as a rule, simply sewn together so as to be large enough to cover, and it is then turned in and tacked down to fit. Felt is bad to match. If it is puckered in when matched and sewn, strain it out, face down, as square as possible on the floor. Then damp the wrinkles, and, with the application of a hot goose, it will shrink to any required shape. All carpets should be well pressed on the seams; but other kinds can not be shrunk to shape by damping and pressing to such an extent as can be done with felt.

Floorcloths, kamptulicons, and linoleums are always fitted at the house with a good sharp knife. It is well to be supplied with an oil-stone, as the edge of the knife is soon gone when cutting such material. In cases where it is to be bordered, the border should be first mitered round the room, not fixing the inside edges. Next put the body material squarely into its place, letting it pass a little under the

border. Mark the body-cloth with a pencil or the edge of a
knife by the inside edge of the border, cut it off just as it
lies on the floor, pull out the cuttings from under the border,
and the edges of body and border will then butt together.
If this is done carefully, an exact fit will be secured. If the
floor is boarded, a few gimp pins will keep the floorcloth in
its place; but if it be of stone a solution should be used
for the purpose of cementing it in its place, which can be
obtained from the makers of linoleum, or good stiff flour
paste might be used. Even when laying kamptulicon on
board floors it is best cemented, as that prevents it stretching and lying loosely.

CHAPTER XI.

CUTTING OF SHADES.

THE cutting of shades may be easily accomplished by taking the length or drop of the shade, as it is called, and the width of the roller, and cutting the shade 12 inches longer than the exact drop. This additional measurement will be taken up by the hem on the bottom, while part of it will be necessary with a view to allow the shade to go once round the roller, for when a wooden roller is shown bare after the shade is drawn down it is an eyesore. It is, therefore, well to have plenty of length. Allow a quarter inch of play for the shade at each end of the shade-roller. The hems on each side of the shade must be raw-edged, and be 1¼ inches wide. Allow a half inch for turning in at the lower end of the shade, then fold 3½ inches up to form the slip for the shade-lath. It is necessary that a shade should run true, and in order to insure this it must be cut perfectly square. The proper way to go about it is as follows: Cut the linen 12 inches longer than the actual drop; fold the material lengthways down the middle; if there be a stripe in the blind, let the fold be exactly parallel with it. Measure across from the fold, both at the top and bottom of the shade, the half width required, and prick it through the double thickness with the regulator; also mark through 1¼ inches beyond; this is to set off the two side-hems. Now mark in the same way a half inch from the bottom; and 3½ inches from that mark up the shade on the outside edge. This gives the turnings and hem for the lath. Also mark a half inch from the top edge to give the line to fix to roller. Unfold the shade, and lay it flat on the board; place the straight-edge from mark to mark, and press a line with the point of the regulator so that the cutting-line will be visible.

It wil then be impossible for the shade to be otherwise than square. It is also a quicker and surer method than using a square. In the illustration (Fig. 45) the small crosses (thus ×) show the pricks of the regulator, and it is the upholsterer's work to fold the hems in their proper place from these pricked marks. Should the bottom be shaped, put an extra slip for the lath just above the shape-line on the room side of the shade. The neatest way to fix a shade is to put one tack at each end of the roller, to wind it round once till it meets, crease it back, and sew it from end to end. This entirely hides the roller should the curtain be drawn completely down, and will never tear away.

The writer's individual opinion is that in the case of the printed shade material now so much in favor, the side that is printed should be put to the window. When the light shines through it, the pattern will be seen very distinctly in the room, whereas the wrong side would look very bad from the street.

Festoon Shades.—A festoon shade, such as is shown in Fig. 46, should be fixed on a piece of wood 3 inches wide, on which the pulley-wheels work in a manner similar to those of a Venetian blind, but no check-action is required. Four festoons will be required to an ordinary window, of say 10 inches each, with a 3-inch margin on the two outside edges. Cut the shade twice the depth to which it is to be finished. On the width allow 4 inches to each festoon; and on the side-margin sufficient to return to the cord-line, which will thus form a 3-inch hem on each side, finishing under the tape. When the hems are made, mark the festoons or tape-lines, and gather the shade up to the depth required, and on these lines sew narrow tapes of the same color as the shade; on the tapes sew small brass rings about 3 inches apart. To weight the shade, a piece of ¾-inch circular bar-iron is covered with the same material and sewn to each

tape at the bottom, just clear of the shaping. The draw-cords
are tied to this rod, and run up through the rings to the
pulleys, as shown in the illustration (Fig. 46). The material

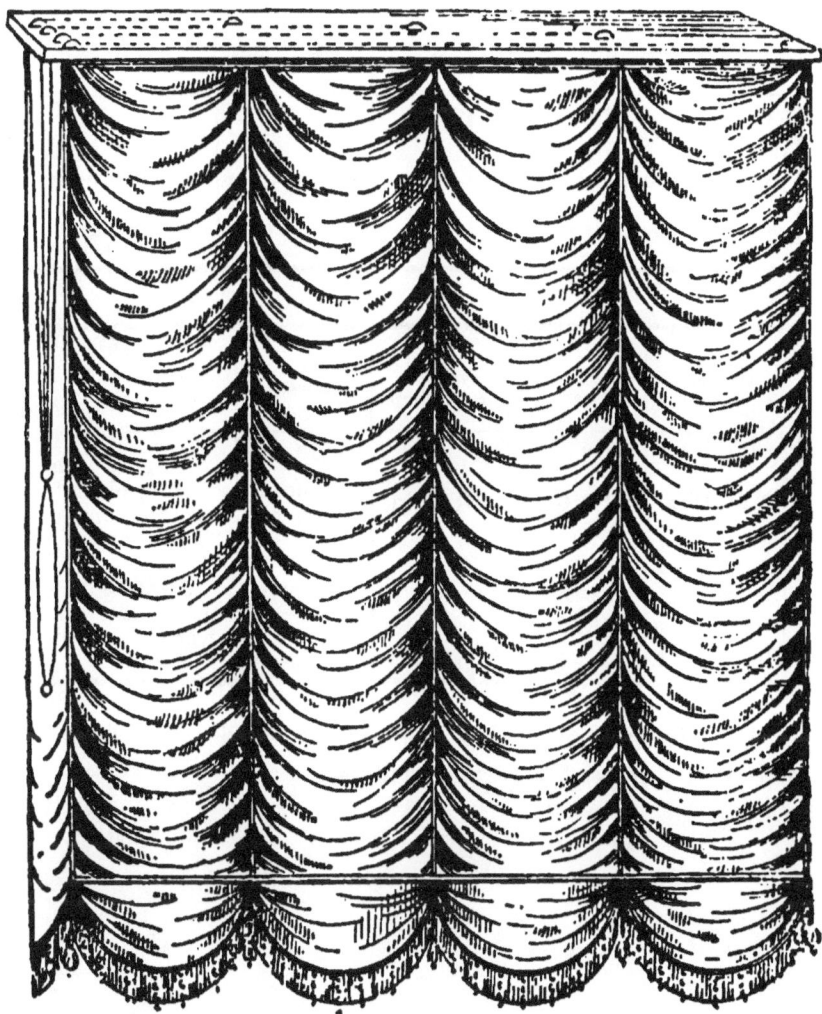

FIG. 46.

used should be thin, or at least soft, and without any dress
or stiffness whatever. The shade is tacked to the top of the
lath before being sewed up, thus hiding the tacks. Put the
tapes to the window in fixing.

It is a good plan to have a book ruled for taking
measurements of shades, something after this style:—

Name, Address, and Date.	Name of Room.	No. of Blinds.	Material. Color.	Description of Roller.
29—1—83 A. Bertrand, No. 1 Pearl.	Library. Bedroom.	Two. One.	No.5, Printed Green Union	Hartserne's Spring.

Width between Beads.	Height between Beads.	Space available for Roller.	How to be fixed.	Sundries. &c.
3 ft., 8 in.	6 ft., 6 in.	4 ft., 4 in.	To head lining.	Straight, with lace, black acorns, brass knot-holders.
4 ft.	5 ft., 6 in.	—	Between beads.	Cleat-hook, knot-holder, and worsted tassel.

By adopting this plan the cost of shades can be esti-
mated, or referred to, with readiness.

CHAPTER XII.

MISCELLANEOUS HINTS.

I THINK the foregoing pretty nearly covers the whole ground of upholstery, though it may be often necessary to apply the rules and designs shown to many other purposes; and I give in this chapter a few designs of things and situations that the upholsterer may sometimes be called upon to deal with.

FIG. 47,

Fig. 47 shows a method of draping the back of a piano. It is quite "the thing" now to have the backs of upright pianos facing visitors, and this fashion renders it imperative to have the backs covered; and nothing adds so much to the artistic richness of the instrument as a harmonious draping in surahs or velours.

Fig. 48 shows another style of draping, which in most cases would be very appropriate.

Figs. 49, 50, and 51 show some odd treatment of French windows that open on a covered porch. The draping on these examples is extremely artistic; and if a proper combination of colors is chosen the effect is very pleasing. As these examples, however, are only given as suggestions, the upholsterer will readily see that the drapery may be changed to an infinite variety of styles.

FIG. 48.

It is true that this branch of the art has fallen into the hands of special artists known as "decorators"; but the two departments are so intimately connected as to be inseparable.

Figs. 52 and 53 show the methods of draping a window from one side only. This style is very effective when there are two windows at one end of a room, and the hangings are gathered from opposite sides.

Sometimes the advice of the upholsterer or his services are required in matters pertaining to general draping, such as the draping of halls, alcoves, niches, and recesses. Much will depend on the taste of the upholsterer, and on the

Fig. 49.

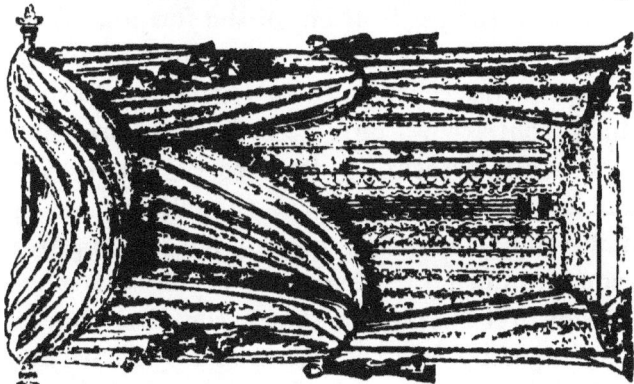

Fig. 50.

Fig. 51.

surroundings of the part to be draped; and all that can be done here to help him out is simply to give a few suggestions and hints.

Fig. 54 shows how some halls are draped when narrow. Wide halls will stand more hangings than narrow ones, and sometimes the amount of fabrics used in the latter is enormous, as may be seen at Fig. 55, which is taken from an actual example.

If we turn to the pages of any of the furniture designers' works of the last century and the first half of the present century, we find descriptions of bed-hangings fearfully and wonderfully contrived,— ugly according to present taste, though probably presenting features of beauty to those who made and used them. Apart altogether from appearance, cumbersome bed-hangings can not be approved of; indeed, some would go the length of saying this about any hangings, however simple, for the reason that they are unhealthy, by impeding free ventilation and by harboring dust. This, no doubt, is a valid objection to the old four-poster, with its heavy hangings carefully drawn to exclude every breath of fresh air from the sleeper, and its tester-cloth to hold the dust from one spring cleaning to another; but, in moderation, bed-hangings such as are now generally seen are not to be decried indiscriminately. Sometimes they may be useful in warding off draughts; and when they are not required for this purpose, they are so light and scanty that they can not be considered injurious. Whether they look well or not is a matter of opinion. They furnish a room, so far as appearance goes, much in the same way that anti-macassars do, as well as being useful in warding off draughts. It is not, however, so much to the hangings as to the upholstering of bed-heads that attention is now directed; for the hangings—that is, the curtains—being principally made up by women, hardly require detailed mention. Of course, I

FIG. 52.

am referring solely to the modern form of bed drapery, which is almost entirely confined to the half-tester style of bedstead, in which, beyond the base-valance, there are only curtains at the head-end. Even, however, in these bedsteads there is considerable scope for display of taste in the upholsterer; but as the varieties of detail may almost be regarded as endless, it is hardly possible to give minute instructions, and these must be confined to the simplest, and, at the same time, fortunately, the most popular style of hangings. But, first, let us see what kind of bedstead is required; and here I may as well say that though wooden bedsteads are by no means things of the past, and will no doubt resume their former popularity, it is still so comparatively rarely they are made now that remarks will be directed to metallic—brass or iron—bedsteads. Of these, one variety—namely, those with head and foot alike, or rather with the head-end a little higher than the foot-end—is not adapted for curtains beyond those from the frame to the floor at the sides and foot. These curtains, if such they can be called, are known as base-valances; and as the only part of their construction that can be considered a man's work consists in the preparation of the laths which are used to connect them with the bedstead, very little need be said about them, as among workers probably ideas of trimming would be scouted by the female amateur who will undertake the necessary sewing. A man, even though an upholsterer, is hardly competent, in the opinion of his "better half," to direct the arrangement of the domestic curtains, though he may design and plan for others. So just leave such little matters to be settled by the feminine portion of the household. You might, however, just hint that to avoid too great depth on the one hand or scrimpiness on the other, the valance should be cut to the measurement between the bottom of the mattress—or, what is the same thing, the top of the frame on

FIG. 53.

which this rests—and the floor; and that for fullness for
pleating, about four breadths of ordinary cretonne width
for a full-length bedstead is a very good standard. I name
cretonne, which, as was stated in one of the preliminary
chapters, is about 31 inches wide, as it is so frequently used
for the purpose; and those who prefer anything else can
easily calculate the number of widths that may be required,
due allowance being made for the thickness of the fabric
and its consequent adaptability for pleating. Roughly
speaking, the length of a base-valance may be about half as
much again as that of the length it is to be when hung, so
that in a 5-foot (wide) bedstead three cretonne widths will
be about right. The base-laths themselves may be made of
pine, a quarter-inch thick and two inches wide. They should
be as long as the side and foot rails on which they rest,—
one of them on each rail. Covering these laths is a long
bag or case, made of the same material as the valance, or
something suitable. The strip of which the base-bag is
formed may be some five inches wide, so that the lath may
easily be inserted in the bag, to which the valance is afterwards regulated and sewn. The palliasse or mattress which
is used will be sufficient to keep the base-lath in position on
the rail; though, should it not do so properly, a few tapes
will make all secure. These, however, are not often required,
if the palliasse is of sufficient size. So much for base-valances,
which, it will be understood, are fixed in much the same
manner to any kind of bedstead; and we may now proceed
to consider the upholstery of head-ends, with a few hints on
the curtains.

Any kind of bedstead may have an upholstered head, by
which, it will be understood, is meant a head-end with an
upholstered panel; but the kind to which this adornment is
generally applied is commonly known as the "Persian." In
this the head-rails are much higher than the foot,— as high,

Fig. 54.

in fact, as the ordinary half-tester, with which it might
almost be confounded by the casual observer, though to
others the differences are readily discernible. The only one
to which attention need be drawn here, as it is sufficient for
all practical purposes at present, is that in the Persian head
the ends run straight up, and do not support any overhanging structure. There are, however, many varieties of
both these and other kinds, which converge so closely on
each other that a slight degree of hesitation may sometimes
be excusable; and, after all, it does not matter much what
the bedstead may be called, for the methods of upholstering
them are much the same, and any one who understands how
to do with the modern "Persian" will have little difficulty
in making whatever adaptations may be necessary by altered
circumstances. First of all, a wooden frame will have to
be made to fill the opening. Now, on looking at an iron
Persian bedstead, as usually made, two thin bands of iron,
or rods, will be seen,— one of them a few inches below the
ornamental upper rails, and the other a foot or two above
the lath-rails or bottom of the bedstead. Each of these
bands will be found to have two or three holes in it; and
from the comparatively rough way in which they are finished,
it will be at once and correctly surmised that they are for
use and not for appearance. The bands are there only for
the purpose of fastening the above-mentioned frame, and
the holes are for the necessary screws. There is, therefore,
no difficulty in arriving at the size of the wooden frame,
which ought to fit loosely in width, and to be long enough
to allow it to be screwed to the stays from behind; for I
dare say it is understood that the frame, when upholstered,
is fastened on the front of the head. In other words, it
must lie *within* the uprights, and on the transverse rails.
Sometimes, when the lower one of these is at a considerable
height above the bottom of the bed, it is necessary to make

Fig. 55.

the frame some inches deeper, in order that there may not
be a space between the bedding and the upholstery. As a
rule, however, bedsteads are turned out by the manufacturer
with due consideration of the upholsterer's needs, and it will
seldom be necessary to make the frame more than six inches
lower than the bottom stay, as the object is merely to prevent a vacant space appearing above the bolster and pillows.
In case of doubt, it is better to have the framing rather
lower than absolutely necessary than that it should be too
short. The frame, which is only required as a support for
the upholstery, is a very simple affair. It may be made of
half-inch pine, three inches or so wide, mortised and tenoned,
halved, or fastened at the corners in any other way that may
be most convenient. It will also be well, especially in the
case of a wide bedstead, to further strengthen it by the
addition of cross and upright stays, which may be either
mortised or halved to the outer framing, the former joint
being, of course, the better of the two. Extreme neatness
of workmanship is not required,— sufficient strength to bear
any strain there may be from the covering being the principal thing to be studied. I shall merely indicate two
arrangements for this, neither of which can be regarded as
novelties in themselves; and perhaps, on that account, will
be more generally useful to the upholsterer than more com-
plicated devices, which, if he wants, he will no doubt be able
to devise for himself. In fact, as I have already hinted, so
much depends on the skill and taste of the worker that it is
almost impossible to give hard and fast lines that must be
adhered to. In the drapery part of the upholsterer's craft
this is specially the case, for a slight difference in a fold, or
some in itself comparatively insignificant detail, may make
all the distinction between a tasteful, artistic arrangement,
and the reverse. Material has something to do with effect;
but, after all, the chief matter is the arrangement, and on

this no merely written instructions can convey ideas suitable for every instance. Color, again, is a very important factor in the appearance of bed or any other furniture drapery, for no hangings can look well unless they are in harmony with the predominant coloring of the room. But this hint being given, it must take care of itself, for any lengthy consideration of coloring would occupy far too much space. Those who want to study this feature thoroughly can not do better than read Chevreul's "Laws of Contrast of Color," a standard work, which gives many valuable suggestions for interior decorative effects depending on color.

Perhaps the easiest, and, at the same time, one of the most effective styles in which the head can be upholstered is that in which the covering is pleated to the center. It looks well in either a plain or figured material. In the latter care must be taken to have the pattern running in one direction, namely, to or from the center throughout. It will not do to have the design on the upper part, starting as it were from the center, and on the bottom part from the lower edge. This applies, of course, only to some of a flowing or floral character, for in designs of a geometrical kind no attention will be required beyond, if the pattern is very large or pronounced, seeing that they match at the joins of the different lengths. To ascertain how many of these are wanted, measure the width of the covering material, and mark this off at intervals on the edge of the frame, not including the angles,— that is, the measurements will ignore the angle. Perhaps this may be made clearer by giving actual measurements. Let us suppose that the size of the frame is 4 feet 6 inches by 5 feet, and the width of the material is 31 inches. Start from any point at the edge of the frame, say the left top corner, and measure 31 inches along the top, then take the same distance from this to the edge of the right-hand side, and so on till the frame has been gone round. The number of lengths required will thus be eight, seven of them being the full width of the cretonne, 31 inches, or whatever the covering is, and the other a mere strip of a few inches. The length of each of these pieces must be equal to the measurement from the center to the furthest portion of the woodwork, allowing for each a fullness of about 5 inches or 6 inches. I should say here that the point where the pleats meet need not be the actual center of the frame, as the object should be to get it about half way between the top of the bedding and of the top of the panel. The remainder are measured in the same way, namely, by taking sufficient length to cover the woodwork to the extreme length. All the ends which are to be in the middle must now be pleated, and the most convenient way to do so is to thread them a short distance from the edge on a piece of wire,— ordinary bell-wire will do very well,— and then bend the wire into a ring, as small as the thickness of the stuff will conveniently allow, say about $2\frac{1}{2}$ inches in diameter for cretonne. The ends of the wire, which should be left long enough for the purpose, must then be passed through a hole bored in the frame at the center and tacked down behind securely. The covering must then be tacked down, being drawn a little way over the back of the frame for the purpose. It then only remains to trim off the superfluous stuff and to cover the center with a rosette, a watch-pocket, or some similar contrivance of an ornamental character. A border may also be added to the panel, which, if covered as directed, will have the pleats very pronounced at the center, and gradually fading into a level surface at the edge.

As the flatness of the edge is merely owing to the widths of the stuff being cut close, it follows that any amount of pleating that may be wanted can be got there by simply allowing more breadths of material. As a rule, it is better to have some pleating at the edges, as the joins of the different widths can be concealed throughout their entire length within the folds. Some little manipulation will be required to get the pleats regular and of equal fullness, for unless this is attended to the effect will be unpleasant. As a guide while fixing, it will be well to mark the spaces within which each piece is to be pleated at the edge. No great amount of skill is required in fixing this kind of head, but neatness is essential. The same may be said of the "hour-glass head."

Bed-curtains.—There is nothing difficult in the making up of bed-curtains, as now usually seen on the half-tester and the Persian with arms, and a little attention to a few simple details will enable any one who can sew to make them up fairly. Of course, there is scope for that quality vaguely termed taste, which every one possesses in abundance, or fancy they do, so that nothing more need be said about it. The materials of which bed-curtains are most frequently made are chintz and cretonne, but any other substance may be used, if preferred. It is as well, however, that it should not be too heavy; and there can be little difference of opinion that the two just named are the most suitable. In an earlier chapter reference was made to the reversible cretonne, which being printed with a pattern showing on both sides, may be made into curtains without requiring any lining. Fabrics printed on one side only are generally, though not invariably, lined. The lining is a thin self-colored fabric sold for the purpose. From a width and a half to two widths are generally used for a curtain which is to be "headed" up to be about 20 inches wide, which is about the average of the Persian arm-bedstead, and may be taken as a standard. The "heading" consists in pleating the top end of the curtain to bring it into pleasing folds when hanging, without the awkward bulging which would result without; though, so far as utility is concerned, curtains which have no fullness are quite as good. Whether they look as well or not is a matter of opinion only. The workman will not require to be reminded that the heading may be formed of either plain box, double box, or running pleat, nor will he require instructions how to stitch them. The front and bottom edges may with advantage be trimmed with fringe binding, the style of which is sufficiently indicated by its name. But these few hints will probably be sufficient for all practical purposes.

Beds, Pillows, Bolsters, Feathers, &c.—The differences between a mattress and a bed having now been stated, it will be readily understood that the construction of the latter is a very simple matter. In fact, it can scarcely be said that, beyond sewing the case and filling it with the requisite materials, there is anything to be done. The case is made of tick, which should be of good substance, to prevent the feathers—the usual filling—from working their way through it. The top and bottom of the case may be sewn to each other, or they may have a border intervening, similar to mattresses, the edges being either bound or welted, as may be preferred. There is an idea held by some people that cases should be waxed inside, but this is by no means necessary nor universal. With very thin loosely woven ticks it may be an advantage, as waxing tends to prevent feathers coming through; but if the ticking is good it is not required. Waxing is open to several objections, besides the obvious one that, if not necessary, it involves a useless expenditure of time and labor. However, as some may think the bed is not complete unless the case is waxed, it may be stated that the operation is a very simple one. A lump of beeswax, the ordinary yellow kind generally, is rubbed all over the *inside* of the case, so that a little adheres,—in my opinion, the less the better, for it forms an admirable holdfast for dust.

Soap may be used in the same way, instead of wax, if preferred. Bolsters and pillows are made in much the same way,—the former being round bags, with pieces shaped accordingly to form the ends; the latter flat, the top and bottom being sewn together without any border. In addition, there is the wedge-shaped variety; but it is so little used in this country that it is hardly necessary to take it into consideration.

Feather beds are often denounced as unhealthy, and, though some may be, they should not all be considered so. A great deal depends upon the way the feathers have been purified and cleansed, for unless this has been thoroughly done beds made of them can not be wholesome. Feathers, to be fit for bedding purposes, must be effectually cleansed and prepared, not only by the removal of dust and loose dirt, but by the destruction of the animal matter contained inside the quills. This can not be done by the old-fashioned rustic plan of hanging them in a bag and beating them occasionally, though now and again the method is advocated as being sufficient for amateur purposes. All that can be said of it is that it is, perhaps, better than nothing. I mention this for the benefit of those who prefer using feathers from their own fowls and poultry, for others will find it better to buy the feathers ready dressed. Perhaps here I may give a hint worth attention by those who keep poultry, namely, that feathers are a marketable commodity, and there is no difficulty in disposing of them in almost any city to purifiers of feathers for bedding purposes. This source of profit from the fowl-run seems to be commonly overlooked in this country. The smaller feathers only should be kept; and where there are a sufficient number of birds to make it worth while, the different kinds should be kept separately. Of course, I do not refer to "moulted" feathers, for the trouble of collecting these would not be compensated for by their value, but to those plucked from the bird when killed, and consequently comparatively clean. The sorts chiefly useful for home cleaning and consumption are ordinary poultry and goose feathers, the latter being far the more valuable of the two. Duck feathers are only permissible in small proportions, and game feathers not at all. It may be well here to note that white feathers are worth more money than gray or darker colors, though for all practical purposes one is as good as the other, the only difference being in appearance. Between poultry and goose feathers there is, however, a great difference in elasticity and filling qualities, goose being by far the best. This renders them more costly,— say about double the value of poultry feathers. I mention these facts for the benefit of those readers who may have a sufficient quantity of poultry and geese to justify them in saving the feathers. Indeed, it is principally for such that the remarks in this chapter are intended.

Poultry and goose feathers may be mixed together, and, if desired, in order to increase the quantity, though at a slight deterioration in quality, the large feathers may also be saved. The feather portion of these can easily be detached from the quill or stalk, which is useless in beds, for very obvious reasons. To render the large feathers usable, it is only necessary to pull them from the stalk, commencing at the outer end, when the soft portion will come away in strips.

Now, it would be very little use giving directions about "home-grown" feathers, unless something was said about cleaning them. This, as done on an extensive scale, could be managed by the amateur purifier, and I can do little more than suggest to him the method he should pursue. Those whose business it is to prepare feathers for bedding purposes, have costly contrivances for washing, drying, &c.,

steam and hot air entering largely into the process. Such appliances naturally are beyond the reach of the workman, who must content himself with simpler plans; but with care they need be little, if at all, less efficacious. Any one who has had experience in manipulating a large quantity of loose feathers will know that though they may be clean things, they are apt to be productive of untidiness; therefore, when working with them, it will be as well to have as little else in the room as possible.

The first thing to be done is to thoroughly wash the feathers, to get rid of external dirt, the amount of which, even on the cleanest feathers, is very considerable. Possibly the idea that feathers may be washed is a new one to some, who when they see them saturated with water, and apparently spoiled beyond redemption, may be inclined to think some mistake has been made. When thoroughly wet they don't look nice, certainly; but never mind, they will be all right when dried, so don't spare the water. Keep changing this till it is no longer discolored by dirt; and if there is any contrivance handy, such as a washing-machine of the ordinary kind, or a dolly-tub, it will be just the thing. If any arrangement can be made by which a constant stream of water can be running through while the feathers are being stirred about, so much the better, as the washing will be expedited. I do not advise that any large quantity should be done at once, for feathers when wet are heavy and apt to clog.

When all dirt is removed,— that is, when the water runs away clean,— in the absence of steam the feathers should be washed further in warm water, in which a small quantity of chloride of lime has been dissolved. Some other disinfectants would do as well; but this being cheap and generally obtainable, besides being one of the best for the purpose, will be the most convenient. The feathers should be well

stirred in this, that it may saturate them all, and left in it
for some time, say till the water is cold. They should next
be washed in water alone to remove the chloride till no
smell remains.

Drying the feathers will probably be the most tedious
part of the undertaking, for they hold a great deal of water,
and must be *thoroughly* dried before they can be slept on.
When dried on a large scale, the best contrivance I have
seen for wringing them is a kind of large sieve or drum, in
which the feathers are put. This drum revolves at a high
rate of speed, and, being perforated, the water is expelled
by centrifugal action. Such a piece of apparatus is, however, costly; and I only mention it by way of a hint for the
amateur feather-dresser to adapt to anything he may have
that seems suitable. When the feathers have been dried as
much as possible by this means, they are still quite wet, the
water they have absorbed still remaining, they are further
dried by hot air. It may be suggested that in the absence
of special appliances, the wet feathers should be loosely
spread on a net raised above the floor, and exposed as much
as possible to sun and warmth. This will take longer, of
course, than the regular plan; but there is no reason, if
space be available, why it should not be adopted. The
feathers, if moved about now and then, will dry by themselves, and the final airing may be given by putting them in
bags of convenient size in a hot room, or before the fire.
Whatever they are put in to dry, remember that feathers
when wet do not occupy anything like the space they
do when dry, and that they should have plenty of room
to expand, as the drying process proceeds. If they are
pressed down too much in drying, their elasticity will be
diminished; and it is on this quality, perhaps, more than
any other that their suitableness as a filling material chiefly
depends. If spread thinly when wet, and allowed to dry

spontaneously, an occasional stirring being given them, their elasticity will take care of itself, but if dried in bags, this must be provided for. It will not do to leave the feathers too long in a sodden state, nor need any fear be entertained of breaking them, and so injuring them by handling them, as when wet they will stand any amount of rough usage. I am hardly prepared to say that the excess of moisture might be squeezed out of them by an ordinary wringing-machine, but I do not think it would do much harm. I should prefer to risk this rather than leave them wet for any extended time, as if left long enough they would be sure to rot; but I dare say means of drying will readily occur to the cleaner, according to the things he may have accessible. Of one thing I would caution the feather-dresser, namely, not to use any bed till the feathers are *thoroughly* dry. He can not be too particular about this, and I emphasize the necessity, as unless he is accustomed to handling feathers, he may think them dry long before they really are. The process, as described, may seem troublesome, and it doubtless is so, but nothing less will suffice to make feathers fit for bedding. Properly purified feathers should be perfectly odorless, not only when cold, but when warmed by the occupier of the bed. It sometimes happens that nothing unpleasant is noticed at first on going to bed, but that in a short time a faint musty odor makes itself perceptible, as the bed or pillow gets warmed by the heat of the body. This betokens that the feathers are not properly purified. Whether such are prejudicial to health I do not pretend to have had any special opportunity of observing, but few will have any doubt on the point. When properly dressed, however, I have little hesitation in saying that feather beds are as wholesome as they are luxurious. Anyway, the present is not the occasion to discuss at length the healthiness or the reverse of feather bedding, and this chapter may be

concluded by giving some idea of the quantity of feathers
required in ordinary feather beds, bolsters, and pillows,
that they may be sufficiently full and yet not too hard.
Something will depend on the elasticity of the feathers,
but the quantities here given may be taken as an average
approximate for good comfortable bedding. In beds, about
8 lbs. per foot in width will be sufficient, taking the length
as for an ordinary full-sized bedstead. Thus, all that it is
necessary to do to ascertain the weight required for any
given width of bed is to multiply the width by 8, which
will give the weight in pounds. For example, a 5-foot (wide)
bed will, according to this scale, give 40 lbs. as the weight
for the feathers. Bolsters should have about 1¼ lbs. per
foot of length; pillows about 2½ lbs. each.

Beds, pillows, and bolsters are occasionally filled with
other substances besides feathers; but as their construction
is much the same, those who wish to do so will have no
difficulty in making them. Perhaps, in connection with
feather beds, I ought to refer to down quilts, now so much
used in place of heavier bed-clothing. The stuffing in them,
though popularly known as eider-down, is really goose-
down. It may be separated from the feathers by fanning
these gently, when, by reason of its superior lightness, it is
blown on one side. It is, however, unpleasant fluffy stuff to
work with; and as nearly the whole of the making of down
quilts is sewing, it will not commend itself to workmen.
Still, there may be some who wish to know how to make
these quilts, and brief instructions may accordingly be given.
The down is put in a cover of some suitable material, such
as satteen, silk, &c., spread evenly, and is then sewn by
"quilting" through.

Carpet-planning.—In a former chapter allusion was made
to carpet-planning. A few more hints, however, may rea-
sonably be given on carpet-planning, cutting, and making.

As is well known, it is not considered necessary nowadays for a carpet to fit all over a floor, the so-called "art-squares," woven in one piece, being preferred by many for both economical and sanitary reasons. They are made in various sizes, and the only thing that need be said is that about 18 inches is a fair width of flooring to allow as a margin. Thus, for a room measuring say 15 feet by 12 feet, the square should be about 12 feet by 9 feet. Of course, these measurements are only approximate, as the "square" carpets are generally obtainable only in certain stock sizes. Occasionally the squares are made up from ordinary widths of carpeting, such as Brussels, Kidders, &c., when they can be worked to any dimensions desired. The measuring for square or center carpets is a very simple matter; but when it comes to making a carpet to fit close or all over a floor, much care is required, and it is rarely that the owner succeeds in giving such measurements as will enable a carpet-planner to cut to an exact fit. I know from experience that the amateur measurer can seldom be convinced that the fault lies with him instead of the carpet-planner. I trust that those who may find that the carpets made to these measurements, but which on being laid do not fit, will excuse me telling them that the cause lies with themselves in almost every instance, and that the carpet-cutter is rarely to blame. I would like to write a lot on carpet-planning, for it would save carpet-dealers much annoyance. Yes, I speak feelingly of that which I do know; but at present I must content myself with saying that any one who measures for carpets must observe the utmost nicety in measuring. First of all, draw a plan of the room; measure each space not merely along the walls, but from angle to angle, not omitting the smallest projection or recess. This plan is afterwards drawn out full size in the carpet-planning room, and the carpet cut to it. The edges, where necessary to prevent fraying, should be turned under and "herring-boned," except, perhaps, at doorways and under fenders, when it is better to finish them with carpet-binding made for the purpose. Laying is a somewhat laborious process, especially when the carpet is new, as it must be tightly stretched before tacking down. I do not recommend the old-fashioned plan of fixing by rings, for, though some prefer it, it is generally better to tack down.

Stair-carpets should be cut full length, so as to allow the position to be changed whenever signs of wear are perceptible on the edges of the treads. As stair-carpets always wear out first at these places, those who have regard for economy will do well to change the lay of the carpet frequently, and for this purpose it is necessary to allow a good surplus. In measuring for stair-carpets, all that is required is to measure the height and the tread of the step, add the dimensions together, and multiply by the number of steps.

Measuring for floor-cloths, whether oilcloth or linoleum, is very similar; but the actual cutting may be done while laying. Floor-cloth should be tacked down, but linoleum is frequently cemented, especially to stone floors. A special cement is made and sold for the purpose; but a very good one, preferred by some to the orthodox preparation, may be found in a mixture of ordinary flour paste, glue, and resin.

Established 1870.

CATALOGUE

OF

STANDARD BOOKS

PUBLISHED AND FOR SALE BY

THE INDUSTRIAL PUBLICATION CO.

9 Barclay Street, New York.

These books will be sent, postpaid, to any address in any accessible part of the world, on receipt of price.

New catalogues, with additions of new books, are issued from time to time, and will be sent free to any address on request.

THE STEEL SQUARE AND ITS USES. — By FRED. T. HODGSON. — Third and Enlarged Edition. Illustrated by nearly one hundred large and clear engravings. Cloth, gilt. $1.00

This is the only really practical work on the steel square and its uses ever published. It is thorough, accurate, clear, and easily understood. Confounding terms and scientific phrases have been religiously avoided where possible, and everything in the book has been made so plain that a boy of twelve years of age, possessing ordinary intelligence, can understand it from end to end.

This new edition, just issued, is illustrated with nearly one hundred handsome engravings, showing how the square may be used for solving almost every problem in the whole art of carpentry. The carpenter who possesses this book need not waste time and material "cutting and trying." He can lay out his work to a hair's breadth, and "cut to the line."

STEEL SQUARES AND THEIR USES. — Being a description of the various steel squares and their uses in solving a large number of mechanical problems in constructive carpentry, joinery, sheet-metal work, cut-stone and brick work. Also showing how many geometrical and other problems may be solved by the use of the steel square. — By FRED. T. HODGSON, editor of "The Builder and Woodworker." — Finely Illustrated. Cloth. $1.00

This forms Part II of "The Steel Square and Its Uses." It gives new problems, new methods, and new wrinkles for shortening work.

.*. With these two volumes in his possession the workman has at command the entire practical mathematics of construction, and is prepared to lay out any piece of work more easily, quickly, and accurately than it can be done by any other method.

PRACTICAL CARPENTRY. — Being a guide to the correct working and laying out of all kinds of carpenters' and joiners' work. With the solutions of the various problems in hip-roofs, gothic work, centering, splayed work, joints and joining, hingeing, dovetailing, mitering, timber-splicing, hopper-work, skylights, raking moldings, circular work, &c. To which is prefixed a thorough treatise on "Carpenters' Geometry."— Illustrated by over three hundred engravings.— By FRED. T. HODGSON, author of "The Steel Square and Its Uses," "The Builders' Guide, and Estimators' Price-Book," "The Slide-Rule, and How to Use It," &c.— Cloth, gilt. $1.00

This is the most complete book of the kind ever published. It is thorough, practical, and reliable; and at the same time is written in a style so plain that any workman or apprentice can easily understand it.

STAIRBUILDING MADE EASY. — Being a Full and Clear Description of the Art of Building the Bodies, Carriages, and Cases for all kinds of Stairs and Steps. Together with Illustrations showing the Manner of Laying Out Stairs, Forming Treads and Risers, Building Cylinders, Preparing Strings; with Instructions for Making Carriages for Common, Platform, Dog-legged, and Winding Stairs. To which is added an Illustrated Glossary of Terms used in Stairbuilding; and Designs for Newels, Balusters, Brackets, Stair-moldings, and Section of Hand-rails.— By FRED. T. HODGSON, author of "The Steel Square and Its Uses," "The Builders' Guide, and Estimators' Price-Book,' "The Slide-Rule, and How to Use It," &c.— Cloth, gilt. $1.00

This work takes hold at the very beginning of the subject, and carries the student along by easy stages until the entire subject of stairbuilding has been unfolded, so far as ordinary practice can ever require. This book, and the following one on "Hand-Railing," cover nearly the whole subject of Stairbuilding.

A NEW SYSTEM OF HAND-RAILING; or, How to Cut Hand-Railing for Circular and other Stairs. square from the plank without the aid of a falling mold.— By AN OLD STAIRBUILDER.- Edited and corrected by FRED. T. HODGSON.— Cloth, gilt, . . $1.00

The system is new, novel, economic, and easily learned. Rules. instructions. and working drawings for building rails for seven different kinds of stairs are given.

HINTS AND AIDS TO BUILDERS.— Hints and Aids in Building and Estimating gives hints. prices. tells how to measure, explains building terms. and, in short. contains a fund of information for all who are interested in building.— Paper. . . . . . . 25 cents

THE BUILDERS' GUIDE, AND ESTIMATOR'S PRICE-BOOK.— Being a compilation of current prices of lumber, hardware, glass, plumbers' supplies, paints, slates, stones, limes, cements, bricks, tin, and other building materials. Also, prices of labor, and cost of performing the several kinds of work required in building. Together with prices of doors, frames, sashes, stairs, moldings, newels, and other machine-work. To which is appended a large number of building rules, data, tables, and useful memoranda; with a glossary of architectural and building terms.— By FRED. T. HODGSON, editor of "The Builder and Woodworker," author of "The Steel Square and Its Uses," &c. 12mo, cloth. $2.00

CARPENTERS' AND JOINERS' POCKET COMPANION. Containing rules, data, and directions for laying out work, and for calculating and estimating.— Compiled by THOMAS MOLONEY, carpenter and joiner.—Cloth. . . . . 50 cents.

This is a compact and handy little volume, containing the most useful rules and memoranda, practically tested by many years' experience in the shop, factory, and building. Also, a treatise on the framing-square. It is by a thoroughly practical man, and contains enough that is not easily found anywhere else to make it worth more than its price to every intelligent carpenter.

EASY LESSONS; OR, THE STEPPING-STONE TO ARCHITECTURE.— Consisting of a series of questions and answers explaining in simple language the principles and progress of architecture from the earliest times.—By THOMAS MITCHELL.— Illustrated by nearly one hundred and fifty engravings.—New edition, with American additions.—Cloth. . . . . 50 cents.

Architecture is not only a profession and an art, but an important branch of every liberal education. No person can be said to be well educated who has not some knowledge of its general principles and of the characteristics of the different styles. The present work is probably the best architectural textbook for beginners ever published. The numerous illustrative engravings make the subject very simple, and prevent all misunderstanding. It tells all about the different styles, their peculiar features, their origin, and the principles that underlie their construction. -

BUCK'S COTTAGE AND OTHER DESIGNS.—Just the book you want, if you are going to build a cheap and comfortable home. It shows a great variety of cheap and medium-priced cottages, besides giving a number of useful hints and suggestions on the various questions liable to arise in building, such as selections of site, general arrangement of the plans, sanatory questions, &c. Cottages costing from $500 to $5,000 are shown in considerable variety, and nearly every taste can be satisfied. Forty designs for fifty cents. . . . 50 cents.
The information on site, general arrangement of plan, sanatory matters, &c., is worth a great more than the cost of the book.

WATER-CLOSETS. — A Historical, Mechanical, and Sanatory Treatise.— By GLENN BROWN, Architect; Associate American Institute of Architects.—Neatly bound in cloth, with gilt title. . . $1.00

This book contains over 250 engravings, drawn expressly for the work by the author. The drawings are so clear that the distinctive features of every device are easily seen at a glance, and the descriptions are particularly full and thorough. The paramount importance of this department of the construction of our houses renders all comment upon the value of such a work unnecessary.

PLASTER: HOW TO MAKE AND HOW TO USE. — Illustrated with numerous engravings in the text, and three plates giving some forty figures of ceilings, centerpieces, cornices, panels, and soffits. Being a complete guide for the plasterer in the preparation and application of all kinds of plaster, stucco, Portland cement, hydraulic cement, lime of Tiel, Rosendale and other cements. To which is added an illustrated glossary of technical terms used by plasterers, with hints and suggestions regarding the working, mixing, and preparation of scagliola and colored mortars of various kinds.—Cloth, gilt.. . . . . . . $1.00

An invaluable book for plasterers, bricklayers, masons, builders, architects, and engineers.

HANDSAWS: THEIR USE, CARE, AND ABUSE. — How to Select and How to File Them.— By FRED. T. HODGSON, author of "The Steel Square and Its Uses," "The Builders' Guide, and Estimators' Price-Book," "Practical Carpentry," &c.— Illustrated by over seventy-five engravings.—Cloth, gilt. $1.00

Being a complete guide for selecting, using, and filing all kinds of handsaws, backsaws, compass and keyhole saws; web, hack, and butchers' saws;—showing the shapes, forms, angles, pitches, and sizes of saw-teeth suitable for all kinds of saws, and for all kinds of wood, bone, ivory, and metal. Together with hints and suggestions on the choice of files, saw-sets, filing clamps, and other matters pertaining to the care and management of all classes of hand and other small saws.

This work is intended more particularly for operative carpenters, joiners, cabinet-makers, carriage-builders, and woodworkers generally, amateurs or professionals.

MECHANICAL DRAUGHTING.—The Student's Illustrated Guide to Practical Draughting.—A series of Practical Instructions for Machinists, Mechanics, Apprentices, and Students at Engineering Establishments and Technical Institutes. — By T. P. PEMBERTON, Draughtsman and Mechanical Engineer.— Illustrated with numerous engravings.—Cloth, gilt. $1.00

This is a simple but thorough book, by a draughtsman of twenty-five years' experience. It is intended for beginners and self-taught students, as well as for those who pursue the study under the direction of a teacher.

HINTS AND PRACTICAL INFORMATION FOR CAB-INET-MAKERS, Upholsterers, and Furniture-men Generally. To-gether with a description of all kinds of finishing, with full directions therefor; varnishes, polishes, stains for wood, dyes for wood, gilding and silvering, recipes for the factory, lacquers, metals, marbles, &c.; pictures, engravings, &c.— Cloth, gilt. $1.00

This work contains an immense amount of the most useful informa-tion for those who are engaged in manufacture, superintendence, or construction of furniture or woodwork of any kind. It is one of the cheapest and best books ever published, and contains over one thousand hints, suggestions, and methods; and descriptions of tools, appliances, and materials. All the recipes, rules, and directions have been care-fully revised and corrected by practical men of great experience, so that they will be found thoroughly trustworthy. It contains many of the recent recipes sold for from $5 to $500.

THE SLIDE-RULE, AND HOW TO USE IT.—This is a compilation of explanations, rules, and instructions suitable for me-chanics and others interested in the industrial arts. Rules are given for the measurement of all kinds of boards and planks, timber in the round or square, glaziers' work and painting, brickwork, paviors' work, tiling and slating, the measurement of vessels of various shapes, the wedge, inclined planes, wheels and axles, levers, the weighing and measurement of metals and all solid bodies, cylinders, cones, globes, octagon rule and formulæ, the measurement of circles, and a compar-ison of French and English measures, with much other information useful to builders, carpenters, bricklayers, glaziers, paviors, machinists, and other mechanics.—Paper.. . . . . 25 cents.

Possessed of this little book and a good slide-rule, mechanics might carry in their pockets some hundreds of times the power of calculation that they now have in their heads, and the use of the instrument is very easily acquired.

THE ENGINEER'S SLIDE-RULE AND ITS APPLICA-TIONS.— A Complete Investigation of the Principles upon which the Slide-Rule is Constructed; together with the Method of its Application to all the Purposes of the Practical Mechanic.— By WILLIAM TONKES. Paper.. 25 cents.

THE LIGHTNING CALCULATOR.— Practical Hints on Lightning Calculating.— To which are added Rules, Tables, Data, Formulæ, and Memoranda for making rapidly those every-day calcu-lations that are required in business, mechanics, and agriculture. Paper.. 20 cents.

HOW TO BECOME A GOOD MECHANIC.— Intended as a Practical Guide to Self-taught Men : telling what to study, what books to use, how to begin, what difficulties will be met, how to overcome them. In a word, how to carry on such a course of self-instruction as will enable the young mechanic to rise from the bench to something higher.— Paper. . . . . 15 cents.

This is not a book of "goody-goody" advice, neither is it an advertisement of any special system, nor does it advocate any hobby. It gives plain practical advice in regard to acquiring that knowledge which alone can enable a young man engaged in any profession or occupation connected with the industrial arts to attain a position higher than that of a mere workman.

LECTURES IN A WORKSHOP.— By T. P. PEMBERTON, formerly Associate Editor of "The Technologist"; author of "The Student's Illustrated Guide to Practical Draughting." With an Appendix containing the famous papers by Whitworth "On Plane Metallic Surfaces of True Planes"; "On the Uniform System of Screw Threads"; "Address to the Institution of Mechanical Engineers, Glasgow"; "On Standard Decimal Measures of Length."— Cloth, gilt. $1.00

We have here a sprightly, fascinating book, full of valuable hints, interesting anecdotes, and sharp sayings. It is not a compilation of dull sermons or dry mathematics, but a live, readable book. The papers by Whitworth, now first made readily accessible to the American reader, form the basis of our modern systems of accurate work.

DRAWING INSTRUMENTS.— Being a Treatise on Draughting Instruments, with Rules for their Use and Care; Explanations of Scales, Sectors, and Protractors. Together with Memoranda for Draughtsmen, Hints on Purchasing Paper, Ink, Instruments. Pencils, &c. Also, a price-list of all materials required by draughtsmen.— By FRED. T. HODGSON.— Illustrated with twenty-four explanatory illustrations.— Paper.. . . . . 25 cents.

PLAIN DIRECTIONS FOR THE CONSTRUCTION AND ERECTION OF LIGHTNING-RODS.— By JOHN PHIN. C.E.— Fully illustrated. 12mo, cloth. . . . . 75 cents.

This is a well-known and standard work. It is the only book on the subject published in this country that has not been written in the interest of some patent, or some manufacturing concern.

CEMENTS AND GLUE.— A Practical Treatise on the Preparation and Use of all kinds of Cements, Glue, and Paste.— By JOHN PHIN, author of "How to Use the Microscope."— Paper. . . . . 25 cents.

Every mechanic and householder will find this volume of almost every-day use. It contains nearly two hundred recipes for the preparation of cements for almost every conceivable purpose.

ELECTRICIANS' POCKET COMPANION.— Electrical Rules, Tables, Tests, and Formulæ.— By ANDREW JAMIESON, C.E., F.R.S.E. 12mo, cloth. . . . . 75 cents.

This is the most compact and thorough work in the market for the practical electrician. It contains minute directions for all calculations, tests, &c., with clear engravings of the apparatus employed. The following list of contents will give an idea of its scope:—
Formulæ of the Absolute Units,— Practical, Electrical, Mechanical. Heat and Light Units.
Electro-chemical Equivalents, Electrolysis, Heat and Energy of Combustion.
Practical Methods of Electrical Measurements.
Electric Conductors: Copper, &c.— Insulators: Guttapercha, &c. Submarine Cables.— Aerial Land-lines.
Electric Lighting, and Transmission of Power.

CENTURY OF INVENTIONS.— An exact reprint of the famous "Century of Inventions" of the Marquis of Worcester (first published in 1663). With Introduction, Notes, and a Life of the Author. With portrait after a painting by Van Dyke.— Edited by JOHN PHIN.— 12mo, extra cloth. $1.00

This is one of the most extraordinary books ever published. The famous "Century of Inventions" is of more than mere historical interest. It contains numerous suggestions and hints of what might be accomplished, and will be found of great interest and value to every one interested in mechanics. Many persons claim that the Marquis of Worcester anticipated many of our most important modern inventions and discoveries. Great care has been taken to reproduce exactly (so far as modern type, &c., could do it) the edition published by the Marquis himself. It has been entirely out of print for many years.

THE NEW POCKET CYCLOPÆDIA.— A Compendium of General Knowledge, Useful and Interesting Facts, Valuable Statistics, and Practical Information. 16mo, cloth. . . . . 50 cents.

This is a handy volume of 164 pages, printed in small but clear type on very fine thin paper, so that the book may be readily carried in the pocket or trunk. It contains all those facts and figures which are most commonly referred to in everyday life, great care being taken to secure accuracy and clearness. It is not a collection of "curious information" made up from newspaper clippings, but a thoroughly arranged manual of the most useful general knowledge.

WOOD-ENGRAVING.— A Manual of Instruction in the Art of Wood-Engraving. With a Description of the necessary Tools and Apparatus, and Concise Directions for their use; Explanations of the terms used, and the methods employed for producing the various classes of wood-engravings.— By S. E. FULLER.— Fully illustrated with engravings by the author, separate sheets of engravings for transfer and practice being added.— New edition. neatly bound. . . . . . . 50 cents.

THE WORKSHOP COMPANION.— A Collection of Useful and Reliable Recipes, Rules, Processes, Methods, Wrinkles and Practical Hints for the Household and the Shop.— Paper, ... 35 cents. Cloth, gilt title, ... 60 cents.

This is a book of 164 closely printed pages, forming a dictionary of practical information for mechanics, amateurs, housekeepers, farmers, — everybody. It contains a series of original treatises on various subjects, such as alloys, cements, inks, steel, signal-lights, polishing materials, and the art of polishing wood, metals, &c.; varnishing, gilding, silvering, bronzing, lacquering, and the working of brass, ivory, alabaster, iron, steel, glass, &c.

THE WORKSHOP COMPANION.— Part II.— This is an extension of the First Part, and contains subjects which have not been discussed in the earlier volume.— Paper, 35 cents; cloth, 60 cents.

These two volumes are not a mere collection of newspaper clippings, like most of the books of "Recipes," but a series of tnorough articles on practical matters in regard to which information is constantly desired in the shop, the house, and on the farm. The two parts are also issued in one volume, printed on extra paper, and handsomely bound in cloth, under the title of "The Practical Assistant."— Price $1.00.

AMATEUR'S HANDBOOK OF PRACTICAL INFORMA-TION for the Workshop and Laboratory.— Second edition, greatly enlarged, neatly bound. . . . . 15 cents.

This is a handy little book, containing just the information needed by amateurs in the workshop and laboratory. Directions for making alloys, fusible metals, cements, glues, &c.; and for soldering, brazing, lacquering, bronzing, staining and polishing wood, tempering tools, cutting and working glass, varnishing, silvering, gilding, preparing skins, &c. The new edition contains extended directions for preparing polishing powders, freezing mixtures, colored lights for tableaux, solutions for rendering ladies' dresses incombustible, &c. There has also been added a very large number of new and valuable recipes.

"The Workshop Companion," just described, contains *all* the matter that is to be found in "The Amateur's Handbook."

COLLODIO-ETCHING.— A Guide to Collodio-Etching.— By Rev. BENJAMIN HARTLEY.— Illustrated by the author.— 12mo, cloth. $1.00

This volume gives complete and minute instructions for one of the most delightful of amateur arts. It is fully illustrated by woodcuts of all the apparatus used (which is very simple and easily made), and also by actual photo prints of the etchings themselves.

HINTS FOR PAINTERS, DECORATORS, AND PAPER-HANGERS.— Being a Selection of Useful Rules, Data, Memoranda, Methods, and Suggestions for House, Ship, and Furniture Painting, Paper-hanging, Gilding, Color-mixing, and other matters useful and instructive to painters and decorators. Prepared with special reference to the wants of amateurs.— By An Old Hand.— Paper. . . 25 cents.

SUCCESS WITH RECIPES.— A Practical Guide to Success in the Use of Recipes, Formulæ, &c.— With Hints on Chemical and Mechanical Manipulation.— Intended as a Supplement to all Books of Recipes.— By JOHN PHIN.— 12mo, paper. . . . . 25 cents.

While it is an undoubted fact that many of the recipes published in the ordinary collections are erroneous, either from original blunders on the part of the authors or from mistakes in copying, failure in the use of others frequently arises from defective information and vicious methods on the part of those who attempt to put them in practice. The object of the present book is to give such hints and cautions as will enable the worker to secure success where success is possible; and where the products are intended for sale it gives special and valuable advice in regard to the best methods of putting them on the market.

TRADE "SECRETS" AND PRIVATE RECIPES.—A Collection of Recipes, Processes, and Formulæ that have been offered for sale at prices varying from 25 cents to $500. With Notes, Corrections, Additions, and Special Hints for Improvements.—Edited by JOHN PHIN, assisted by an experienced and skillful pharmacist.— Cloth, gilt title. Price, ... 60 cents.

This work was prepared by the author for the purpose of collecting and presenting in a compact form all those recipes and so-called "trade secrets" which have been so extensively advertised and offered for sale. It is not by any means a claptrap book, though it exposes many claptraps. It contains a large amount of valuable information that can not be readily found elsewhere; and it gives not only the formulæ, &c., for manufacturing an immense variety of articles, but important and trustworthy hints as to the best way of making money out of them. Even as a book of recipes it is worth more than its price to any one who is interested in the subject on which it treats.

WHAT TO DO AND HOW TO DO IT IN CASE OF ACCIDENT.— A book for everybody.— Cloth, gilt title. . . . . 50 cents.

This is one of the most useful books ever published. It tells exactly what to do in case of accidents, such as severe cuts, sprains, dislocations, broken bones, burns with fire, scalds, burns with corrosive chemicals, sunstroke, suffocation by foul air, hanging, drowning, frost-bite, fainting, stings, bites, starvation, lightning, poisons, accidents from machinery and from the falling of scaffolding, gunshot wounds, &c. It ought to be in every house, for young and old are liable to accident, and the directions given in this book might be the means of saving many a valuable life.

COMMON SENSE IN THE POULTRY-YARD.—A Story of Failures and Successes. Including a full account of 1,000 hens and what they did. With a complete description of the houses, coops, fences, runs, methods of feeding, breeding, marketing, &c.; and many new wrinkles and economical dodges.— By J. P. HAIG.—With numerous illustrations.— 12mo, cloth, gilt. $1.00

A lively and entertaining work, which embodies the actual experience of many years in the keeping of poultry in large and small numbers. It is the most thoroughly practical work on poultry in market.

A DICTIONARY OF PRACTICAL BEE-KEEPING.—Notes and Practical Hints.—By JOHN PHIN, author of "How to Use the Microscope," &c.—Numerous illustrations.—Cloth, gilt. . . . 50 cents.

This work discusses thoroughly nearly five hundred subjects. Gives in condensed form an immense amount of valuable information under the different headings. Under the heads *Bee, Comb, Glucose, Honey, Race, Species, Sugar, Wax,* and others, it brings together a large number of important facts and figures which are now scattered through our bee literature and through costly scientific works, and are not easily found when wanted. Here they can be referred to at once under the proper head.

SHOOTING ON THE WING.—Plain Directions for Acquiring the Art of Shooting on the Wing. With Useful Hints concerning all that relates to Guns and Shooting, and particularly in regard to the art of Loading so as to Kill. To which has been added several Valuable and hitherto Secret Recipes of Great Practical Importance to the Sportsman.—By An Old Gamekeeper.—Illustrated.—12mo, cloth. . . . . 75 cents.

This book contains a novel and most valuable feature which is found in no other work on this subject. This is a series of graduated lessons by which the self-taught young sportsman will be enabled to advance step by step from such marks as a sheet of paper nailed on a fence to the most difficult trap-shooting and the sharpest snap-shots.

THE PISTOL AS A WEAPON OF DEFENCE in the House and on the Road.—12mo, cloth. . . . . 50 cents.

This work aims to instruct peaceable and law-abiding citizens in the best means of protecting themselves from the attacks of the brutal and the lawless, and is the only practical book published on this subject. Its contents are as follows:—

The Pistol as a Weapon of Defence.—The Carrying of Firearms.—Different Kinds of Pistols in Market.—How to Choose a Pistol.—Ammunition, different kinds; Powder, Caps, Bullets. Copper Cartridges, &c.—Best Form of Bullet.—How to Load.—Best Charge for Pistols.—How to regulate the Charge.—Care of the Pistol: How to Clean it.—How to Handle and Carry the Pistol.—How to Learn to Shoot.—Practical Use of the Pistol.—How to Protect yourself and how to Disable your Antagonist.

CHEMICAL HISTORY OF THE SIX DAYS OF CREATION.—By JOHN PHIN, author of "How to Use the Microscope."—12mo, cloth. . . . . 75 cents.

In this volume an attempt is made to trace the evolution of our globe from the primeval state of nebulous mist, "without form and void," and existing in "darkness," or with an entire absence of the manifestation of the physical forces, to the condition in which it was fitted to become the habitation of man. While the statements and conclusions are rigidly scientific, it gives some exceedingly novel views of a rather hackneyed subject.

THE SUN: A Familiar Description of his Phenomena.— By Rev. THOMAS WILLIAM WEBB, M.A., F.R.A.S., author of "Celestial Objects for Common Telescopes."— With numerous illustrations.— 12mo, cloth. . . . . 40 cents.

This work gives in a delightfully popular style an account of the most recent discoveries in regard to the sun. It is freely illustrated.

HOW TO USE THE MICROSCOPE.— A Book of Practical Hints on the Selection and Use of the Microscope. Intended for beginners.— By JOHN PHIN, editor of "The American Journal of Microscopy."— Sixth edition. Greatly enlarged, with over eighty engravings in the text, and eight full-page engravings, printed on heavy tint paper. 12mo, cloth. $1.25

This work has been received with such general favor that it has passed through five large editions in a few years. It gives a full account of the different kinds of microscopes; of the various accessories, and of the best methods of using them; of the best methods of collecting. preparing, and preserving objects, and preparing slides and cabinets. Many of the illustrations, devices, and methods used, are original with the author, although they have been freely copied and appropriated without credit by several other writers.

A BOOK FOR BEGINNERS WITH THE MICROSCOPE. Being an abridgement of "How to Use the Microscope."— By JOHN PHIN.— Fully illustrated, and neatly and strongly bound in boards.— Price, ... 30 cents.

This book was prepared for the use of those who, having no knowledge of the use of the microscope, or. indeed, of any scientific apparatus, desire simple and practical instruction in the best methods of managing the instrument and preparing objects.

THE MICROSCOPE.— By ANDREW ROSS.— Fully illustrated.— 12mo, cloth, gilt title. . . . . 75 cents.

This is the celebrated article contributed by Andrew Ross to "The Penny Cyclopædia," and quoted so frequently by writers on the microscope. Carpenter and Hogg, in the last edition of their works on the microscope, and Brooke in his treatise on Natural Philosophy, all refer to this article as the best source for full and clear information in regard to the principles upon which the modern achromatic microscope is constructed. It should be in the library of every person to whom the microscope is more than a toy. It is written in simple language, free from abstruse technicalities.

**MICROSOPE OBJECTIVES.— The Angular Aperture of Microscope Objectives.— By Dr. GEORGE E. BLACKHAM.— Eighteen full-page illustrations, printed on extra fine paper.— 8vo, cloth. . $1.55

This is the elaborate paper on Angular Aperture read by Dr. Blackham before the Microscopical Congress, held at Indianapolis.

SECTION CUTTING.—A Practical Guide to the Preparation and Mounting of Sections for the Microscope, special
prominence being given to the subject of Animal Sections.—By SYL-
VESTER MARSH, M.D.—Reprinted from the London edition.—With
illustrations.—12mo, cloth, gilt title. . . . . 75 cents.

This is undoubtedly the most thorough treatise extant upon section
cutting in all its details. The American edition has been greatly enlarged by valuable explanatory notes; and also by extended directions,
illustrated with engravings, for selecting and sharpening knives and
razors.

MARVELS OF POND LIFE.—A Year's Microscopic Recrea-
tions among the Polyps, Infusoria, Rotifers, Water-Bears, and Polyzoa.
By HENRY J. SLACK, F.G.S., F.R.M.S, &c.—Seven full-page plates, and
numerous wood engravings in the text.—Second edition.—12mo, cloth,
gilt. $1.00

DIATOMS.—Practical Directions for Collecting, Preserving, Trans-
porting, Preparing, and Mounting Diatoms.—By Professor A. MEAD
EDWARDS, M.D., Professor CHRISTOPHER JOHNSTON, M.D., Professor
HAMILTON L. SMITH, LL.D.—12mo, cloth. . . . . 75 cents.

This volume contains the most complete series of directions for collecting, preparing, and mounting diatoms ever published. The directions given are the latest and best.

BACTERIA.—A Series of Papers on the Exhibits at the Biological
Laboratory of the Health Exhibition, under the charge of WATSON
CHEYNE.—Reprinted from the London "Lancet."—Illustrated with
over thirty engravings, showing the forms and modes of growth of the
various species, and the apparatus used in the different "cultures."
12mo, paper. . . . . 25 cents.

**HANDBOOK OF URINARY ANALYSIS, Chemical and
Microscopical.**—For the use of Physicians, Medical Students, and
Clinical Assistants.—By FRANK M. DEEMS, M.D., Laboratory Instruct-
or in the Medical Department of the University of New York; Member
of the New York County Medical Society; Member of the New York
Microscopical Society, &c.—Second edition. Greatly enlarged, and
fully illustrated.—12mo, cloth. $1.00

This manual presents a plan for the systematic examination of liquid
urine, urinary deposits, and calculi. It is compiled with the intention
of supplying a concise guide, which, from its small compass and tab-
ulated arrangement, renders it admirably adapted for use, both as a
bedside reference-book and a work-table companion. The author is
well known as one who has had for several years a very extended experience as a teacher of this important branch of physical diagnosis, and
he has compiled a manual which will serve to lessen the difficulties in
the way of the beginner, and save valuable time to the practitioner

THE MICROSCOPIST'S ANNUAL FOR 1879.—Contains a list of all the microscopical societies in the country, with names of officers, days of meeting, &c.; alphabetical and classified lists of all the manufacturers of microscopes and objectives, dissecting apparatus, microscopic objects, materials for microscopists, in Europe and America, &c.; postal rates, rules and regulations, prepared expressly for microscopists. Weights and measures, with tables and rules for the conversion of different measures into each other; customs duties and regulations in regard to instruments and books; value of the moneys of all countries in United States dollars; value of the lines on Nobert's test-plates; table of Moller's probe-platte, with the number of lines to inch on the several diatoms, &c.; focal value of the objectives of those makers who number their objectives (Hartnack, Nachet, &c.); focal value of the eye-pieces of different makers; magnifying power of eye-pieces and objectives. &c. The whole forming an indispensable companion for every working microscopist.—Limp cloth, gilt. . 25 cents.

TRICHINÆ SPIRALIS: How to Detect Them and How to Avoid Them.—A Popular Account of the Habits, Modes of Propagation, and Means of Dissemination of Pork-worms or Flesh-worms.—By JOHN PHIN, author of "How to Use the Microscope."—Fully illustrated.—Paper. . . . . 25 cents.

A BOOK ABOUT BOOKS; or, Practical Notes on the Selection, Use, and Care of Books.—Intended as a popular guide for book-buyers, students, and all lovers of good reading.—Cloth, 40 cts.

This is a readable, gossipy book, full of literary anecdotes, and containing also a great deal of practical information, useful to every one that owns or expects to own books. The directions for binding, repairing, preserving, and handling books should receive the careful attention of every one that desires to keep his books in good condition; but even if regarded as mere pleasant reading for a leisure hour, there are few more readable and interesting books than this with its gossip, chat, and stories.

It is illustrated with three full-page engravings, one being a reproduction of the first wood engraving of which there is any record; the second is an exceedingly curious woodcut representing the birth of Eve, and the third is an engraving of one of the curious "hornbooks" of the seventeenth century.

RHYMES OF SCIENCE: WISE AND OTHERWISE.—By OLIVER WENDELL HOLMES, BRET HARTE, INGOLDSBY, Prof. FORBES, Prof. J. W. McQ. RANKINE, Hon. R. W. RAYMOND, and others.—With illustrations.—Cloth, gilt title. . . . . 50 cents.

THE YOUNG SCIENTIST.—This Journal was devoted to amateur Science and Art, and was deservedly a favorite with young people. We have a few bound volumes for sale. Price 75 cents each.

THE AMERICAN JOURNAL OF MICROSCOPY.—A few bound volumes for sale. Price $1.00 each.

FIVE HUNDRED AND SEVEN MECHANICAL MOVE-MENTS.—Embracing all those which are most important in Dynamics, Hydraulics, Hydrostatics, Pneumatics, Steam Engines, Mill and other Gearing, Presses, Horology and Miscellaneous Machinery; including many movements never before published, several of which have only recently come into use.— By HENRY T. BROWN, editor of the *American Artisan.*— Eleventh edition.— 12mo, cloth. $1.00

This work is a perfect cyclopedia of mechanical inventions, which are here reduced to first principles, and classified so as to be readily available. Every mechanic that hopes to be more than a mere hewer of wood and drawer of water ought to have a copy.

EASY EXPERIMENTS IN CHEMISTRY AND NATURAL PHILOSOPHY.— For Educational Institutions of all Grades, and for Private Students.— By G. DALLAS LIND, author of "Methods of Teaching in Country Schools," and "Normal Outlines of the Common School Branches."— Paper. . . . . 40 cents.

This book, besides being a valuable guide for the teacher and the student, will afford scientific amusement sufficient to brighten the evenings of a whole winter.

WORKSHOP RECEIPTS.— For the use of Manufacturers, Mechanics, and Scientific Amateurs.— The most complete and reliable collection published.

First Series.
Bookbinding; Candles; Drawing; Electro-Metallurgy; Engraving; Gilding; Japans; Photography; Pottery; Varnishing, &c.— 450 pages, with illustrations. $2.00

Second Series.
Industrial Chemistry; Cements and Lutes; Confectionery; Essences and Extracts; Dyeing, Staining, and Coloring; Gelatine, Glue, and Size; Inks; Paper and Paper-making; Pigments, Paint, and Painting, &c.— 485 pages. $2.00

Third Series.
Alloys, Electrics, Enamels, and Glazes; Glass, Gold, Iron, and Steel; Lacquers and Lacquering; Lead, Lubricants, Mercury, Nickel, Silver, Tin, Vanadium, Zinc, &c.— 480 pages, 183 illustrations. $2.00

Fourth Series.
Waterproofing; Packing and Storing; Embalming and Preserving; Leather Polishes; Cooling Air and Water; Pumps and Siphons; Desiccating; Distilling; Emulsifying; Evaporating; Filtering; Percolating and Macerating; Electrotyping; Stereotyping; Bookbinding; Straw Plaiting; Musical Instruments; Clock and Watch Mending; Photography, &c.— 495 pages. $2.00
.*. In ordering single volumes be particular to mention the "series" wanted.

Any person sending an order (accompanied with the cash) for any two of the four series will receive, without extra charge, a copy of "*The Practical Assistant,*" described on another page.

IN PRESS.

FACTS AND FIGURES FOR PRACTICAL MEN.—Containing Data, Formulæ, Rules, Tables, and Calculations used by Architects, Blacksmiths, Bricklayers, Builders, Cabinetmakers, Carpenters, Contractors, Engineers, Farmers, Firemen, Founders, Gunsmiths, Joiners, Machinists, Masons, Metalworkers, Millwrights, Painters, Plasterers, Roofers, Roadmakers, Surveyors, Tinsmiths, &c.—By JOHN PHIN, author of "Practical Treatise on Lightning-Rods," "How to Use the Microscope," &c.

It is intended to make this work a convenient and reliable manual for those who have no knowledge of the higher mathematics. It will be published in four parts, price 25 cents each. Part I is nearly ready.

READY.

THE PRACTICAL UPHOLSTERER.—This work contains a number of original designs in drapery and upholstery, with full explanatory text and an immense number of working illustrations.

It gives a description of tools, appliances, and materials. Tells how to upholster chairs, parlor furniture, bedroom furniture, &c. It contains rules for cutting bed-hangings, window-curtains, door-hangings, and blinds, and for measuring and cutting carpets. Gives arithmetical calculations for cutting carpets, curtains, &c., mantelboard drapery, festoons, and, in short, everything pertaining to upholstery.

There is nothing published in this country that is so thorough and complete in the instructions given for upholstering as this book.

12mo, handsomely bound in cloth. Price $1.

JUST PUBLISHED.

THE PRACTICAL POCKET COMPANION.—A Cyclopedia of the most useful Facts, Figures, and General Information required by Everybody in Everyday Life. To which is added a Concise Dictionary of the most important new terms recently introduced into Science and the Arts.—Edited by JOHN PHIN, author of "How to Use the Microscope," "The Workshop Companion," "Chemical History of the Six Days of Creation," &c.—Numerous illustrations.—Price 20 cents.

In Preparation.

THE UNIVERSAL

CARPENTER AND JOINER.

By FRED. T. HODGSON,

Author of "The Steel Square and Its Uses," "Practical Carpentry,"
· "Stairbuilding Made Easy," "The Builder's Guide, and
Estimator's Price-Book," &c.

To be handsomely illustrated, and published in four parts,
as follows:

PART I will contain CARPENTER'S GEOMETRY, presented in an easy
form, with examples of its practical applications in Carpentry
and Joinery, showing how to get and work
difficult "cuts" and pitches.

PART II will be an introduction to the ART AND SCIENCE OF CAR-
PENTRY proper, with working drawings of Roofs,
Bridges, Bevel and Skew Work.

PART III will contain a practical treatise on AMERICAN JOINERY,
with descriptions and examples of some of the best work
in existence. This Part will be full of excellent
material for the Joiner and Finisher.

PART IV will contain a large number of examples in mixed JOINERY
and CARPENTRY, Rules for Working, Methods of Working,
Tables, Data for Estimating, Formulæ for laying
out work, and Recipes useful to the
Practical Workman.

It will be the aim of the author to make the above work the most
complete treatise on Carpentry and Joinery published to date. It will
be issued in Four Parts, each Part consisting of not less than 100 large
royal octavo pages, printed on fine paper, and strongly bound in stiff
paper covers. The price will be one dollar per Part; and when complete the whole will form one volume, the price of which, handsomely
bound in extra cloth, will be $5.

The First Part will be ready about the first of August, 1891, and the
remaining Parts at intervals of about three months.

To those sending advance orders for two Parts, accompanied with
the cash, we will send a coupon which will be accepted by us as payment
in full for binding the complete work in neat cloth, with gilt title.